THE ALBRECHT PAPERS
VOLUME II
Soil Fertility and Animal Health

ABOUT THE AUTHOR

Dr. William A. Albrecht, the author of these papers, was Chairman of the Department of Soils at the University of Missouri College of Agriculture, where he had been a member of the staff since 1916. He held four degrees, A.B., B.S., M.S., and Ph.D., from the University of Illinois. During a vivid and crowded career, he traveled widely and studied soils in Great Britain, on the European continent, and in Australia.

Born on a farm in central Illinois in an area of highly fertile soil typical of the cornbelt, and educated in his native state, Dr. Albrecht grew up with an intense interest in the soil and all things agricultural. These were approached, however, through the avenues of the basic sciences and liberal arts and not primarily through applied practices and their economics. Some teaching experience after completing the liberal arts course, with some thought of the medical profession, as well as an assistantship in botany, gave an early vision of the interrelationships that enrich the facts acquired in various fields when viewed as part of a master design.

These experiences led him into an additional undergraduate and graduate work that was encouraged by scholarships and fellowships until he received his doctor's degree in 1919. In the meantime, he joined the research and teaching staff at the University of Missouri.

Both as a writer and speaker, Dr. Albrecht served tirelessly as an interpreter of scientific truth to inquiring minds, as this volume amply illustrates. Dr. Albrecht retired in 1959 and passed from the scene in May 1974 as his 86th birthday approached.

THE ALBRECHT PAPERS

VOLUME II

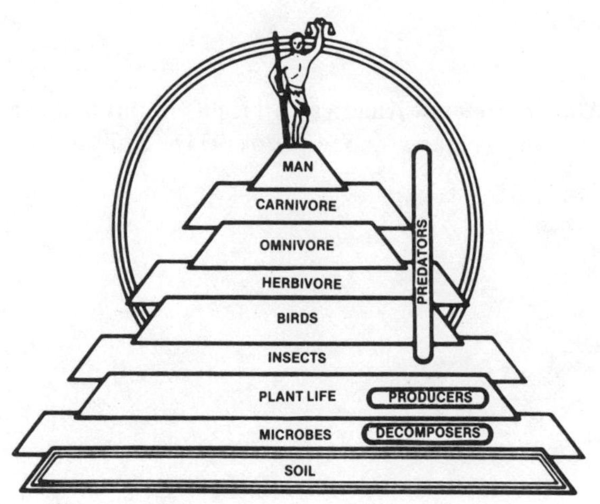

Soil Fertility and Animal Health

by William A. Albrecht, Ph.D.

Edited by Charles Walters, Jr.

Dedicated to the memory of Granville F. Knight, M.D.

CONTENTS

FOREWORD

Among the collection of documents known as The Albrecht Papers *are handwritten notes, notes by students and miscellaneous paragraphs, unsigned, penned in the distinctive hand of Dr. William A. Albrecht. Sometimes the notes of students are most interesting.*

Here, for instance, are notes taken by Russell Leonardson of a meeting on February 21 and 22, 1964, during which Dr. Albrecht explained to Brookside consultants something about what it takes to bring the lessons of the classroom to the field.

"The soil is the creative material of most of the needs of life. Creation starts with a handful of dust," Albrecht told the consultants. "Unless you use nature's evidence in most of your decisions in the field, you may be stampeded into an inaccurate solution. Do not let pride or slipshod observation stampede your decisions."

"You are going to learn only as you are the humble teacher of this man, your client, and as you both work together. Do not try to solve the problems all alone. You need to work together. Remember, he is your graduate student. Always remember that you are the teacher, helping the farmer to resolve his own particular problems and understanding how to work together with nature. It is knowing what your potential is and knowing exactly what your farmer's problems are that makes the program click."

Any Albrecht lecture, any Albrecht paper had its own reason for being. Through a long and crowded career, the great scientist remained a servant with integrity and uncommon common sense. "Do not ask the seed's pedigree to substitute for good nutrition," he often said.

A number of Dr. Albrecht's papers would serve well for this foreword. The publisher has selected one published by the Victory Farm Forum, and reprinted by the Chilean Nitrate Educational Bureau, circa 1948. Styled Quality of Crops Also Depends on Soil Fertility, *it is included here as a perfect prelude for those who will read this volume on* Soil Fertility and Animal Health. *All the papers published here, with the exception of the foreword entry and the epilogue,* Good Horses Require Good Soils, *first appeared in the* Angus Journal, *and are being recycled here with the permission of the publishers.*

Here, then, are Dr. Albrecht's own words from Quality of Crops Also Depends on Soil Fertility. *They, in turn, set the stage for the 13 chapters that follow.*

It is a familiar advertising slogan that says "The quality will be remembered long after the price is forgotten." Implied in this are the simple facts that price is only an arithmetical number; it fixes itself in our minds readily; it is quantitatively cataloged; and its relations and implications are clearly understood. Quality, however, is not so simply measured, not so instantly recognized, and not so completely cataloged. The appreciation of quality is more often a matter of time and experience with the goods. The importance of quality grows on us slowly and thereby gives us time to forget the price while the conviction grows that quality is, after all, the real value. Concerning the quality of our agricultural products in relation to the fertility of the soil, a much stronger conviction is yet to be expected.

In dealing with the soil as the basis of production, yields in bushels, bales, and tons per acre have long been the orthodox yardstick. These are quickly ascertained, easily tabulated, and readily subjected to mathematical manipulations. However, when the productivity of the soil pushes itself into making those bushels and tons more directly our own human food, then quality takes its place as the foremost criterion by which soil should be evaluated. In agricultural production, we are slowly coming to consider the higher quality in our foods and feeds grown on fertile soils as more important than our transitory reactions to questions of price only. We are beginning to see declining soil fertility causing declining quality in our agricultural products despite our struggles to maintain their quality and price. Quality as a criterion for agricultural (food) production is taking on increasing importance.

Land as an economic asset is measured in acres. Soil as the foundation of the agricultural industry is also measured too commonly in the same two dimensions—length and breadth. Soil as the agricultural creative capacity must include a third dimension, namely depth. Now that erosion, by removing surface soil in the humid areas, has exposed much of the infertile subsoil there, we have gained rather forcibly a new appreciation of the productive quality of the upper horizon of the soil profile.

Yield varies both as quantity and also as quality according to the depth of the surface soil. Shallow surface soils are considered "droughty." We are prone to believe that a shortage of water is responsible for the lowered yields of corn on thin soils even when the roots are taking water from the moist subsoil. The real trouble is that the roots are not finding there the fertility required. Bushels per acre are reduced. So is the feeding value per bushel and ton of both grain and fodder.

Quality of the crop does not vary directly with the quantity. Quality drops lower and faster often than the quantity. It may also improve faster than the quantity increases. An increase in the depth of the surface soil does not necessarily represent merely a corresponding increase in capacity for quantity production, while doubling the depth of surface soil may double the yield, it may improve quality even more.

The mass of the soil is not an indication of the fertility adsorbed on the clay or of that held as nutrient reserves in the mineral other-than-quartz making up the skeleton of the soil. Nor is the soil mass an index of the active organic matter with its dynamic values rising and falling as the cyclic temperature of the season controls its behavior to make it correspond more closely with the demands of the growing crop. Quality of the soil, that is its well-balanced supply of fertility, is beginning to stand out in our minds more prominently than do land dimensions either as acres or as depth of the soil. Soil conservation is also coming to be recognized as something more than keeping the body of the soil at home on the farm. It involves, also, the development of a deeper, more fertile and therefore, a stronger soil body.

Crop quality in its relation to soil fertility may have different properties and values in our minds. There is, however, no confusion about crop quantity. In the mind of the baker, quality of wheat refers to its gluten, that is protein content. This same quality in the mind of the farmer has been characterized as the "hardness" of the wheat. The miller, too, is interested in this quality because he makes the flour purchased by the baker. As the purchaser of wheat from the farmer, and as seller of it to the baker, he too is concerned with this quality. The volume and the weight of the loaf of bread can be larger from the use of less flour according as there are more of these qualities centering around the protein in the wheat and in the flour made from it. The miller and the baker, then, search the market for quality first because it determines their volume of output and margin of profit. For them, quality comes first and quantity second.

For the farmer, the volume of output is important too. But his volume is a quantitative matter of bushels of wheat per acre. Even though there may be some premium for "harder" wheat, this extra earning per acre by higher quality is small in contrast to the greater monetary grain he can make from more bushels per acre.

It is quality that is wanted by the baker who uses the flour as he looks to the market offering him his necessary goods. It is quantity that is wanted by the farmer as he looks to the same place to purchase from him the goods he has for sale. Unfortunately, the consumer and the producer in this case do not meet each other. Nor do their separate desires meet through the common market. Both are content to serve the market rather than each other. Consequently, when lowered fertility in the surface soil layer makes many bushels of "soft" wheat with its low protein, the farmer sees his desires well satisfied by the market. He is then blind to, and unconcerned about, the desires of the miller and the baker on the other side of the market. They, too, do not see the declining quality of the soil—especially its declining nitrogen supply—as the reason for the increasing quantity of "soft" wheat in place of "hard" wheat of the high quality they desire. They may accuse the farmer of growing a poor variety, selected for quantity production when, in fact, he had been compelled by declining soil fertility to use it in consequence of which his quality declined while quantity increased.

The quality of corn, like that of wheat, is also premised on soil fertility. We have recently seen increased production not only because of higher rainfall in the western fringes of the cornbelt—from which increases in the national corn crop usually come—but also from the increased yields per acre in the cornbelt proper. At the same time, livestock feeders have been clamoring for help in providing more protein supplements. Corn producers and consumers have been oblivious to the fact that while there had been a tremendous increase in quantity, there has been a serious decrease in quality amounting as protein to almost one-tenth of the total production. It has been reported that during the last ten years, the protein content of corn has fallen from an average of near 9.5 to 8.5%. Fattening power in the form of the starch of the grain has increased. Growth power in the form of protein has decreased. Consequently, while steers feeding on the corn may do well in laying on fat, they may not be building a muscular and bony structure strong enough to carry the extra weight to the market.

Starch production, with its fattening and fuel values, calls for little soil fertility. It calls more for air, water, and sunshine to fabricate this energy-providing food substance. Protein production, however, calls for nitrogen, calcium, and many other items from the soil in addition to the carbohydrates within the plant before it consumes much of its own supply of this photosynthetic product while it converts a part of it into the different amino acids of which the life-carrying and body-growing compound of protein consists. Soil fertility is in reality the foundation of quality in the cereal crops. But, unfortunately, while watching the bushels of wheat and of corn increase to our economic satisfaction, we have been slow to see the decrease in quality or in protein that determines life itself.

While cereals vary some in their quality according to the fertility of the soil growing them, the legume crops in their failures or successes vary more widely. We are learning rapidly that insufficient soil fertility shows greater deficiencies in the legumes which may adversely affect the quality of the animal products and even the animals that consume them.

As we proceed in the biotic pyramid, that is from soil to plant to animal, the effects of deficiencies in soil fertility assume larger proportions. Oblivious of this fact, we have not generally appreciated the variable feed quality of legume hays and forages according to the fertility of the soil growing them. We have recognized the greater difficulty in growing legumes than cereals, but have not been so quick to see the seriously low nutritional value in legumes when they are barely surviving and when we minister to them by means of only one kind of soil treatment. When a single treatment like superphosphate is used, the lespedeza crop, for example, seems much improved if judged on tonnage alone. But when the hay is fed to sheep, these animals make only a fair gain. More significant, however, the animals fail to grow a wool that will last through the scouring process and that can be carded and used in the manufacture of fabrics.

Quite different is the nutritional quality of the lespedeza when the soil is given both limestone and phosphate. The yield as tonnage is increased only slightly. But when fed to the sheep, their growth rate is increased while the wool is heavily loaded with yolk and comes out of the scouring process as readily cardable fibers fit for fabric manufacture. This physiological product of the animal reflects the difference in soil quality more pronouncedly by its widely variable quality than does the animal, the legume, or the cereal in the declining order of irregularities in these different products dependent on the soil.

Here, then, the deficiency of calcium of the soil reflects itself in the cereals, but dimly. It reflects itself more prominently in the legumes; quite decisively in animal growth, but disastrously in the animal products. This chain of differences showing increasing magnitudes with distance from the soil holds in similar manner for deficiencies of copper in the soils of Australia. Shortage of copper there may not seriously disturb the yield of wheat, it may give only a suggestion of its deficiency in animal behavior; but it will be clearly evident in the "steely" wool which the inspector quickly recognizes and discounts on its poor quality.

In this series of products with increasingly specific decreases of quality because of the soil deficiency—going from soil to cereals to legumes, to the animal body and then to the animal products—we are moving from mainly carbohydrate products and photosynthetic activities toward more protein and biosynthetic performances as the major part of our concern. Decreasing quality, then, in this apparent distance from the soil, points to magnification of the soil deficiencies in nitrogen and other fertility elements connected with all the life processes centering in the protein molecule. Quality in agricultural products is calling for more soil fertility by which to make proteins rather than bulk, the promiscuous but simple chemical composition of which permits its production on many soils where protein production of significant food import is almost an unknown.

When the old world has been pointing out to us that its food problem is one of protein rather than of calories, and when war rationing made us conscious of the same thing in our own food problem, shall we not move rapidly to believe that high quality in cereals, in legumes, and in animal products call for more fertile soils to increase in these their concentrations of proteins and minerals as well as the quantities of them? To date, quantity has been our criterion of production even to the extent of detailed statistical treatment of the simple figures reporting them. But good nutrition, fecund reproduction and what we call good health in any form of life do not depend on the quantity, but rather on the quality of the products consumed. Such quality in the final analysis must carry through from the soil as the starting point of the creation of quality as well as quantity.

CHAPTER 1

Nature's Soil Pattern
for Animal Nutrition and Health

"ALL FLESH IS GRASS" were the words by which a prophetic pre-Christian scholar revealed his vision of how the soil, by growing the crops, can serve in creating animals and man. It duplicates to a fairly good degree any concepts we have even now of the many natural performances in the assembly line which starts with the soil to give what we call agricultural production. We know that the soil grows grass; that the grass feeds our livestock; and that these animals in turn, as meats, are our choice protein foods. Along the same thought line we may well consider the geological, the chemical, the biochemical and the biological performances by which the numerous streams of life take off from the soil and continue to flow through the many healthy species of plants and animals. We can, therefore, connect our soil with our health via nutrition. Since only the soil fertility, or that part of the soil made up of the elements essential for life, enters into the nutrition by which we are fed, we may well speak of animal health as premised on the soil fertility.

That the health pattern of animals should be a result of the pattern of soil fertility is the suggestion from authors also of the Roman and Early Christian era reporting their awareness of the fact that plants are no healthier than the plants which nourish them. Not quite so long ago in Great Britain, Izaak Walton, in his *The Compleat Angler*, pointed out that the soil fertility is a factor in determining the quality of sheep wool, and in the tastiness of the trout. "And so I shall proceed to tell you," he says, "it is certain that certain fields in Leominster, a town in Herefordshire, are observed to make sheep that graze upon them more fat than the next, and also to bear finer wool; that is to say, that in that year in which they feed in a particular pasture, they shall yield finer wool than they did that year before they came to feed upon it, and coarser again if they shall return to their former pasture; and again return to a finer wool, being fed on the finer wool ground. Which I tell you that you may better believe that I am certain, if I catch a trout in one meadow he shall be white and faint, and very likely be lousy; and as certainly as if I catch a trout in the next meadow, he shall be strong and red and lusty and much better meat; trust me, scholar, I have caught many a trout in a particular meadow, that the very shape and enameled color of him was made such as hath joyed me to look on him; and I have then with much pleasure concluded with Solomon, 'Everything is beautiful in its season!' "

In that observation of 300 years ago, Izaak Walton saw (more clearly than in just a vision) the differences in the health, in the wool, in the quality of fiber, in the sheen of the body color, in the quality of the muscle meat, and even in the presence or the absence of insect infestations of the beast and the fish in the fields and streams, all related to the fertility of the soil. Hence our discussions to follow, as parts in the broader subject of *Soil Fertility and Animal Health*, are in reality an old and long familiar theme to the keenly observing naturalists, even though to us as scientists it may seem still new, strange and not entirely proved.

DISEASES AS DEFICIENCIES AND DEGENERATION OF THE BODY

There are increasing reports that animal afflictions are coming to be viewed as sins of omission, more than that our livestock is falling prey to some stealthy force (possibly microbes). There is a growing interest in the relationship between their nutrition and their diseases. It is obvious that if an animal is deprived of some essential body constituent, troubles in its health will be

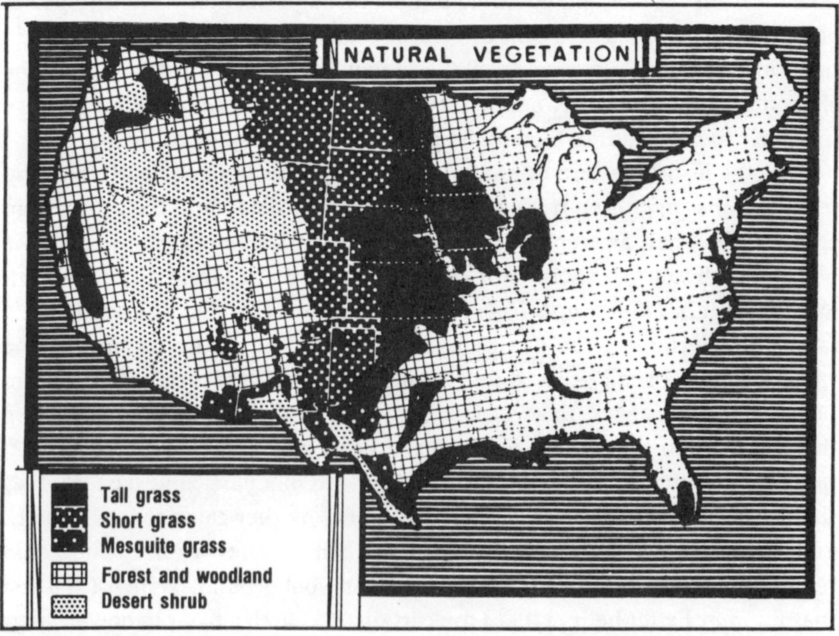

Nature placed the virgin grasses in the midcontinent, and our livestock, like the grass, is balanced along the 98th meridian of longitude, west.

encountered. One may look for that trouble primarily in the direction in which the particular constituent functions. For instance, if a young animal is starved for protein, *i.e.*, the flesh-forming constituent, the result will be a retarded growth replaced possibly by only a fattening; a failure to protect against invasion and digestion by foreign proteins (microbes, viruses, etc.) and a lack of fecund reproduction, which three functions of many compounds are still unknown, and the failing functions as reasons for the disturbed health are too often unrecognized.

The supplies of carbohydrates, fats, and proteins, have received attentions as to their quantities considered necessary for nutrition, but their qualities are not fully tabulated. For the carbohydrates, the latter is not so serious since as energy sources, or body fuels, they in their variety seem safely interchangeable. The fats in an extensive array, also as energy foods, seem to have only four specifically considered to date as indispensible. But for the proteins, too long considered in the group as a whole under this term, must now be viewed as to their quality by which they provide eight or ten specific amino acids as parts of the proteins molecule. These are considered the indispensibles among nearly two dozen of them by which in total their functions are not fully understood.

Only recently, however, as a relatively new phase of the science of animal nutrition, has concern arisen about the supplies of the inorganic constituents of the animal body. When it has been reported that animals deprived as com-

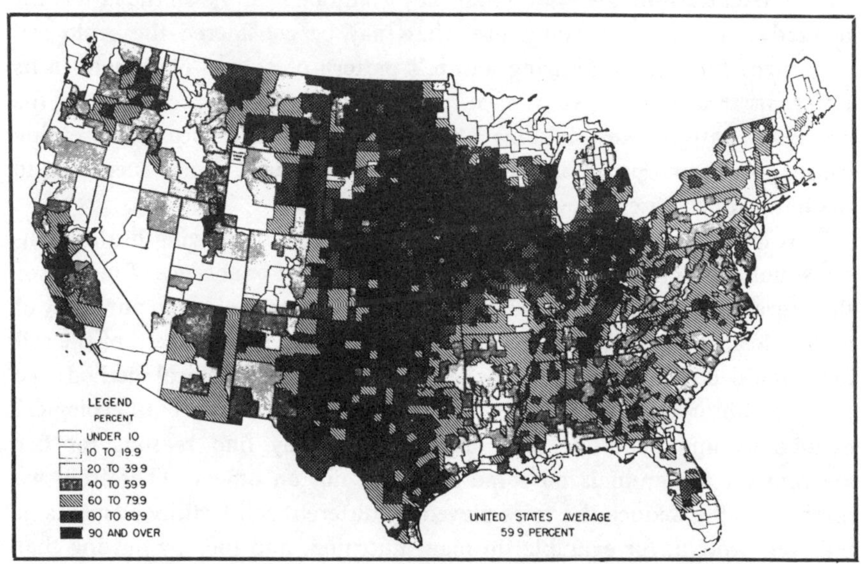

The highest concentration of land in farms, like the virgin grass and livestock today, also is balanced along the 98th meridian of longitude, west.

pletely as possible of the mineral elements in their food died sooner than animals deprived of food altogether [*Dairy Goat Journal*, July 1955], we can see the significance of the growing knowledge of the functions in nutrition of the inorganic elements. These include the "trace" elements as the central cores of the organic enzymes, by which many of the biochemical reactions of the body are propagated. We are recognizing the importance of the balance of the inorganic or "ash" elements more recently. Thereby, we are coming to see, via their source, the importance of the soil for the nutrition of microbes, plants, animals and man. It is the soil, then, that is the foundation of that entire biotic pyramid, and may well be studied in relation to health now that the knowledge of soil is organizing itself into one of the newer and none-the-less significant sciences. As such, soil science may well assume its responsibilities regarding possible contributions to better health for that purpose.

LOOK TO MOTHER NATURE FIRST

In dealing with this subject of the significance of the fertility of the soil as it feeds (or fails to feed) our animals into good health, we can use two methods of approach, either the inductive or the deductive one. In the former, *i.e.* the inductive method, we would first try and learn how each element or factor in soil fertility renders nutritional services to the plant and through that, in turn, to the animal under experiment. Then from that collection of data and experience we would piece together and tell the final story about managing soils to feed the animals so well that they would be healthy. In the latter, the deductive method, we would use what may be considered the ecological approach. In this, by studying nature's pattern of animal placement in its different areas, or on soils of different fertility levels according to the ecological pattern, we would learn the soil fertility differences representing causes, via the crops, of animal presence and animal absence according to the natural processes of evolution of them.

This latter is a qualitative attack on the problem of growing healthy animals, not a quantitative one. It notes the presence of or absence of health, or the prevalence of certain tabulated ailments. We may well begin our study of animal health in relation to the soil fertility and our management of the soil for that objective by using mainly the deductive or the ecological method. We may well observe and investigate nature's pattern. Then, by both ecological deductions and experimental inductions, we may find reasons for her locating certain animals on some soils and not on others. Therefrom we might possibly deduce the roles played by different soil fertility elements in building protein, for example, in plant nutrition, and thereby putting that into animal nutrition. We might thereby learn more about why "All flesh is grass."

As a basic premise from which to reason, or to attack the problem of growing healthy animals by judicious management of the soil under the forages feeding them, we shall accept the American bison's presence in great herds upon the plains as evidence of good fertility in the soil there for healthy animals. We shall consider it as a good fertility for growing the virgin forages of the chemical composition representing nutrition for good animal health. It avoids the distorted view of nutrition which is so often serving mainly in a fattening process, and even then of a castrated male with a much shortened life span which is demonstrating little of a nutrition for the procreation and survival of his species.

By selecting within the larger soil pattern, as the starting point and guide, this particular section which was of sufficient fertility to guarantee the survival of the bison, one is immediately impressed with the limited grass areas which made the buffalo flesh. Or conversely, one is disturbed by the large land areas of virgin soils of fertility levels too low, or too deficient in at least one or two respects, for survival of this quadruped, which is not widely different from the cow. We should also be impressed by the applicability of these ecological facts to the nutrition and health of our cattle herds and other

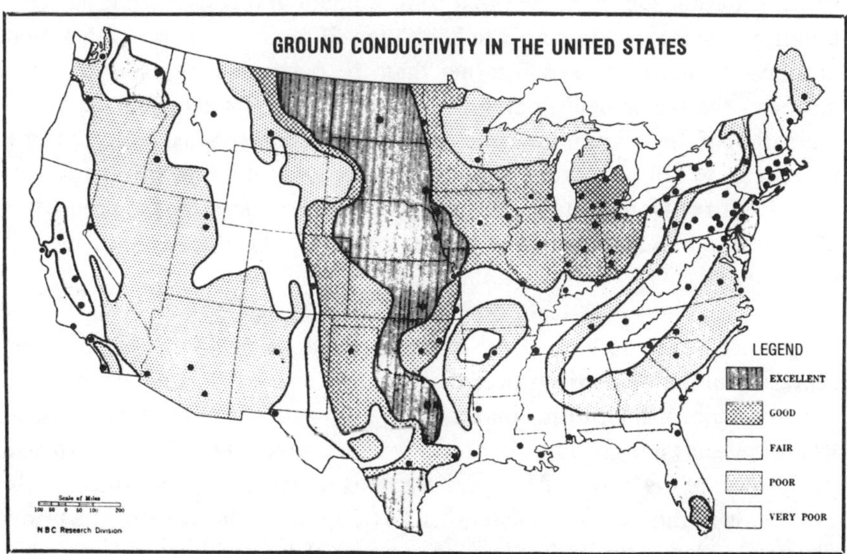

The "excellent" and "good" conductivity of the soil for corresponding radio reception in the United States, like the virgin grass and our livestock, also balance along the 98th meridian of longitude, west. The maps on the preceding pages suggest that (a) the grass, (b) the concentration of farms, and (c) the radio efficiency are not causing each other. Rather the soil fertility, with its electrodynamics resulting from rock weathering, is the common cause in a very specific pattern of control.

livestock when the bison nourished on those soils was duplicating, in many details, their physiological complexities. These include the physiological processes of body growth, of protection against disease, and of reproduction in sufficient fecundity for survival of the species without the help (or hindrance) of man and his modern, economic and technological agriculture. When the soil area suited for raising the bison was so limited, or rather when so much of the country could not support the buffalo, should we not expect to have health troubles with our animals when we put them on soils outside the virgin grass area inhabited by the buffalo? Should we not anticipate more of those troubles with increased years during which those soils have been mined of their fertility?

Let us then examine the concept that the grass crop was the guarantee of the survival of the bison. We may well also look with question into the belief that a grass agriculture anywhere else in the country should be good nutrition and good health for cattle, when they, like the bison, are ruminants and meat-makers too. Let us look at the areas of our own country as the livestock practices are meeting up with more diseases, or less of them, to suggest that the different climatic settings have made different soils from the original rocks or parent soil materials; that some climatic force, like wind, has been blowing in some new minerals to keep the soil reserves renewed for protein production; and that the soil rather than the particular crop species per se has been the factor determining the health of the animals. Perhaps an examination of the soil as it has been developed under the varied climatic forces to feed, or to fail, our protein-producing crops will not only give pattern to the soils of the United States, but will also be the pattern for the production of our cattle and other kinds of livestock in better health.

THE MIDCONTINENT STANDS OUT

One needs only to look at the maps of the states drawn to the sizes they represent in the amounts of beef they produce and in the amounts of pork, our two major meats, to see that it is the midcontinent of the United States where the soils are growing the feeds by which these animals are produced. The growth of the beef which represents a higher percentage of protein in the carcass, is in the western midcontinent. It duplicates the areas of the bison. The production of the pork is located in a part of the midcontinent extending itself eastward over Iowa, northern Missouri, Illinois, Indiana and Ohio. These areas balance well to the east and to the west along the 98th meridian of longitude as the midcontinent line from north to south.

If we are given to the belief that "All flesh is grass," then the readiness with which the maps of beef and pork superimpose themselves so accurately on the area of the virgin grasses would tempt one to believe that the grass causes the ecological and climatic setting for the livestock. We have long considered cattle and grass a "natural" combination in the midwestern area where the grass is still "natural," and where the virgin grasslands and the bison were a "natural" combination too. There the cattle raise themselves. There they are "naturally" fed and are "naturally" healthy. Hogs, however, might seem ill placed on the grass map were one not reminded that corn too is one of the grasses. It grows now where the massive amounts of big bluestem covered the prairies. Hogs, then, occupy the eastern projection of the grasslands, or that extended point of the midcontinent.

It is a significant coincidence that the combined two maps of beef and pork production cover the same area represented also by the portion of the United States with the highest concentration of farms on the land area. We might be tempted to believe that the original grasslands were reason for the productivity making much soil tillage or farming and thereby much livestock possible, or vice versa. These combinations of maps center the attention on the midcontinent as the area where agriculture in terms of protein production is at a high level and correlated with grass, many farms, and much livestock.

CORRELATIONS ARE NOT NECESSARILY CAUSES

It is significant that the maps of beef and pork superimpose themselves also accurately on the map of highly efficient conductivity of electricity by the soil, especially with the combined areas of "excellent" and "good" radio reception as mapped by the National Broadcasting Company. While in the preceding correlations we might mistake the grass as reason for many cattle, and they in turn as reason for many farms, none of these can be mistaken as reason for the excellent and good radio reception, or the high conductivity by the soil.

Quite the reverse, the soil conditions representing "excellent" and "good" conductivity correlated with grass, cattle and concentration of farming may properly be considered as the cause within the correlation. The former one is the cause of the latter three. The three, namely grass, livestock and concentration of farms, are related because they have a common cause, namely the electrochemodynamic conditions of the soil which are exhibited in the radio and also in the production of the protein-producing crops serving in the nutrition of animals and man.

These soil conditions include chemically active and thereby electrically active salts in the soil for the conduction of electricity when the soil is one arm of the radio circuit while the air with its electrical waves is the other. Those conditions include also moisture, since dry salts are neither ionized nor chemically active in electrical conduction. Those are the same soil conditions required for plant nutrition when the salts include those of calcium as the major portion of the mixture, supplemented by those of magnesium, potassium and others.

They are the conditions including and emphasizing water. Thus, with the higher rainfall in the eastern and southeastern states where the soil fertility elements have been leached out, there is ample water but not enough of the fertility salts for either good radio reception or for production of protein-rich crops. There the *virgin pine trees are the extreme of low-protein crops.* In the more arid west, the alkaline salts are ample but the moisture is deficient for the electro-dynamic behaviors by which either good radio reception or significant protein production by crops may be possible. It is in the salts of soil fertility and the soil moisture as the result of the balance of the climatic forces for this particular degree of soil development, that we find the cause of the radio reception, of the natural grasses, of the high concentration of farms and of the possibility of creating the healthy life forms in our livestock and also of man. Underneath all these are the fertility dynamics of the soil. Without these arranged in nature's pattern we cannot expect to grow healthy cows.

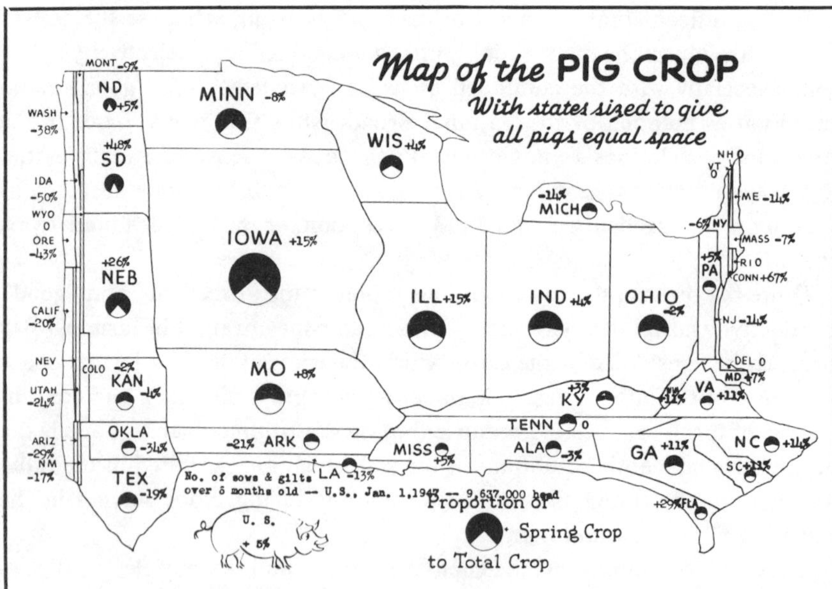

Number in each state shows percent increase or decrease in number of sows and gilts over 6 months old on January 1, 1947 from pre-war years 1939-41. Prepared by American Meat Institute. Source of Data: U.S.D.A. [This pattern has not changed greatly since World War II.]

CHAPTER 2

The Climate Makes
the Soil Pattern

SINCE THE CATTLE, sheep and hogs of the United States locate themselves in their larger numbers in the midcontinent according to the soil pattern, it is but natural to raise the next question, namely, "What makes the soil pattern?" A map of the pig concentration shows this fattened and short-lived animal dominating the eastern tip of the country's central area in which most of the livestock grows. The long-lived, high protein-producing beef cattle have pushed themselves farther west in that mid-region. In making a traverse from its eastern tip to the western edge of our major livestock territory, which is also the area of (a) virgin grass, (b) higher concentration of farms, and even of (c) more efficient radio reception, one is reminded by a look at the map of annual rainfall that this area ranges from only moderate to low precipitation. It is, then, subject to shortages of soil water for production of large yields of vegetative bulk in forage crops.

RAINFALL AND TEMPERATURE DETERMINE
THE DEGREE OF SOIL DEVELOPMENT

The fact that the virgin grass, which made many bison, should locate itself and feed this kind of virgin livestock in the areas subject periodically to severe drought, emphasizes a major illustration of the great fact that the climatic forces of rainfall and temperature determine the degree of development of the soil. Then this, by means of its stock of fertility accordingly, determines what vegetation grows as nutrition and, thereby, what lives in a given area. This great fact of ecology is further illustrated by the absence of forest trees in that location, save those equipped to cut their leaf surface back to the water supply during drought and located mainly in the drainage basins with a bit more soil moisture.

Trees require a continuous supply of soil water during the season for their growth. They cannot go dormant for a period and then start growth again during the season in accordance with the interruption and resumption of the soil moisture supply. The growing tissue of the trees is at the extremities or ends of their branches. In the grasses, that is within the heart or core of the grass clump. It is nearer the soil and the source of water and nutrition, as it were. Consequently this is the systematic safety factor through which a drought may wither the top growth and the grass may die back to let the plant become dormant, but will return to growth again during the same season when the soil moisture favors.

Grass is, therefore, a kind of vegetation specially fitted, through the course of its evolution, for survival in climatic settings of the rainfall and temperature illustrated by the midcontinent and the western states (excluding the coastal ones). There the annual rainfall decreases by longitudinal belts in going westward toward the Coast Ranges. In that great area, the rainfall is scarcely ever the equivalent of the annual evaporation from a free-water surface. It is not high enough to bring about much soil development from the original rocks and minerals. Instead, the soils may be calcareous, saline or even alkaline according as the precipitation-evaporation ratio (as a percent) goes farther below 100 or as the evaporation is that much greater than the annual rainfall on going westward from the midcontinent.

By going in the opposite direction, or starting from the Coast Range then, the increasing rainfall and corresponding increase in development of soils brings us to those from which the alkalis and excessive salts have been washed out. In the midcontinent they are well stocked with calcium (and

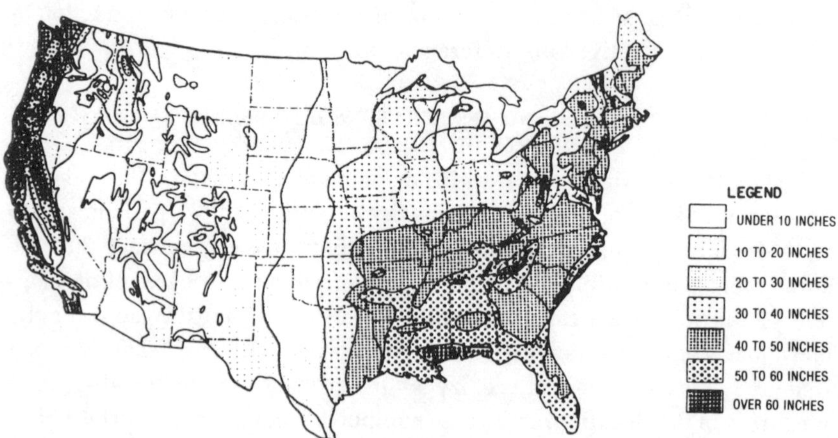

It is the climate that makes the soil by which microbes, plants, animals and man are nourished. The amount of rainfall gives a pattern with increasing degree of soil development in going from the arid west to the humid east of the United States.

magnesium) instead of much sodium. They may have a layer of calcium carbonate which was leached out of the surface soil and deposited in the subsoil. This caliche layer is at increasing depths in the soil profile, and in decreasing amounts of thickness on coming nearer the midline of the midcontinent, namely the 98th meridian of longitude. Calcium and magnesium represent the main store of active nutrients in these soils. All the other essential elements, requiring no more leaching for removal than does calcium, are also part of that fertility store. In that combination, nature provides a decided diversity since droughty conditions with winds from the drier west favor their transportation and deposition of minerals, or a winnowing service, to guarantee the more silty surface soils and an extensive array of nutrient essentials for legumes and their active nitrogen fixation. These facts served to establish a mixed herbage of high-mineral and high-protein concentrations and to give deep, dark soils of excellent structure with a generous stock of nitrogenous organic matter in them under virgin conditions.

MODERATE RAINFALL MEANS SOIL CONSTRUCTION FOR PROTEIN PRODUCTION

It was the moderate climatic forces bringing about the rock decomposition at only a moderate rate that resulted in a particular kind of clay residue, and

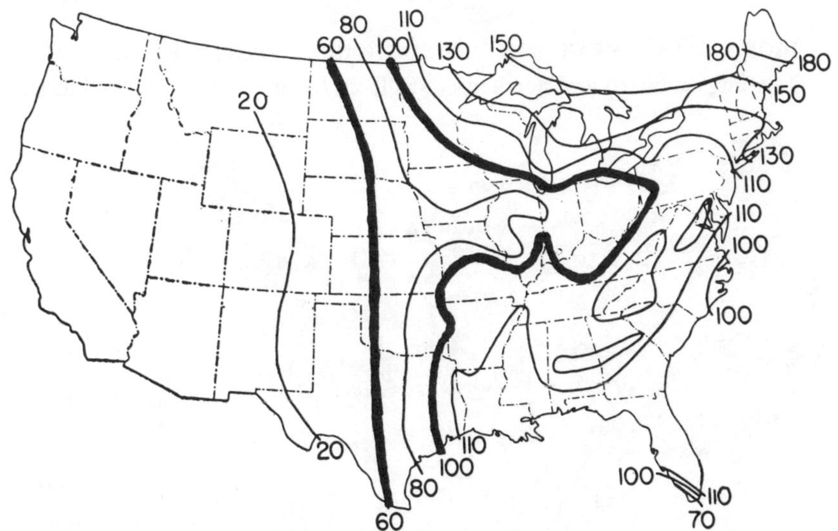

Distribution of Transeau's precipitation-evaporation ratio in the United States (80).

The temperature modifies the effects of the rainfall in developing soil from the rocks, according as the annual evaporation is greater (in the west) or less (in the east) than the corresponding precipitation. The midcontinent emphasizes itself for its well, but not excessively developed, soils. They have ample clay, generous fertility, and good reserve of unweathered minerals.

at the same time stocked the clay's absorbing power with this extensive lot of nutrient elements ranging from the major to the trace ones. Soils so stocked had no room for hydrogen or acidity. These conditions left plenty of unweathered minerals as reserves for later decomposition. They removed the excess sodium, or the alkalinity and salinity. They made the region a good supply and balance of the inorganic parts of the soil fertility through only moderate rainfall and moderate temperature. Those climatic forces, along with the soil resulting therefrom, emphasize the soil and not the climate directly as the high potential for protein production by the herbage. This serves the plant for its own better survival through nitrogen fixation, and then, by its own death, the addition of the nitrogen to the soil.

Nature in the raw is seldom mild. Each species of life, whether microbe, plant, animal or man, is struggling to get its protein supply. Hence the deep, dark colored soil, its high contents of nitrogen and organic matter, and its production of mineral-rich, protein-rich feeds to grow the bison and the beef cattle are all manifestations of climate forces of soil construction, or those developing these particularly favorable soil conditions from the original rocks and their minerals. As the climate gave us the soil, so the soil, in turn, gave the nutrition for healthy animals.

HIGH RAINFALL MEANS SOIL DESTRUCTION FOR PROTEIN PRODUCTION
On going eastward and northeastward from the midcontinent, the increas-

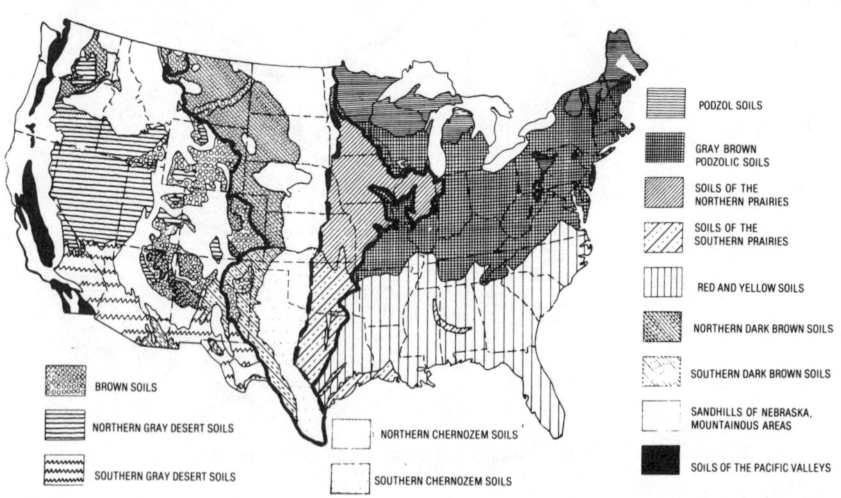

On the soil map the prairies with their grasses, the deep black soils known as chernozems, and the dark brown soils to their west, all help us see the midcontinent as the livestock area because the climatic forces make the fertile soils as the underlying cause of protein-rich feeds.

ing rainfall means more soil development, even to the high degree of soil destruction in terms of protein production. The increasing ratio of rainfall to evaporation in that traverse means more water going through the soil to leach it. Also, the higher rainfall growing more vegetable bulk means more of this to decay, with more carbonic acid resulting to have its hydrogen or acidity replace the soil's calcium and magnesium. In such climatic areas and on soils so highly developed, the protein-rich vegetation, like the grasses, does not result. It cannot obtain the required fertility for such quality production. Instead, the forests making mainly carbohydrate, like cellulose, prevail there. The agricultural crops like corn are less highly nourished for protein production by soils developed under the higher rainfall and are crops that naturally produce mainly carbohydrates also. By feeding these, the castrated males of either cattle or hogs are readily fattened. These soils so highly developed, under the rainfalls high enough to provide plenty of water for large crop

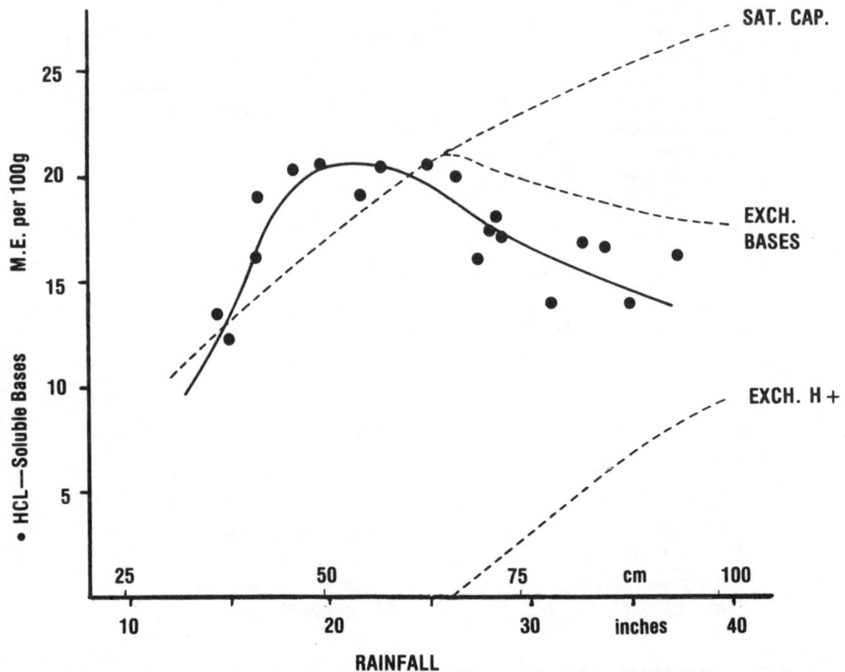

Tests of the soils, along a line of constant temperature but increasing rainfall (going from west to east), show increasing saturation capacity of the soil with more rainfall. The fertility as exchangeable nutrients increases up to the 25 to 30 inches of rainfall in the midcontinent but decreases with higher rainfalls as hydrogen or acidity comes into the soil to replace them.

yields, represent soil destruction so far as protein production is concerned. Their virgin hardwood and coniferous forests gave testimony of this fact.

Moderate rainfalls in the west and high rainfalls under moderate temperatures in the northwest decompose the rock to form a clay of much capacity to hold or absorb nutrients in the west, but one which holds mainly hydrogen, or acidity as a non-nutrient, in the northeast. Thus while the soils of these eastern states with this kind of clay may be highly acid, they are nevertheless potentially good soils. Were we able to apply fertilizers to stock them with the same array of nutrients, as were in the western soils when growing the bison, we would have protein-producing soils for healthy cattle

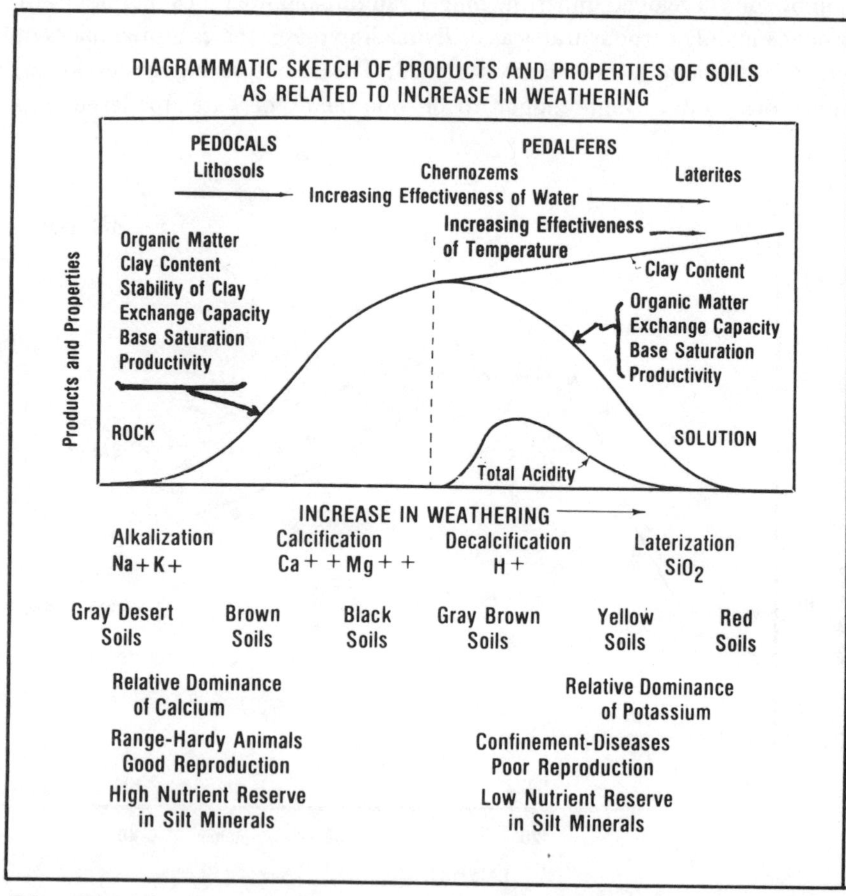

Diagrammatic representation of the development of soil under increasing forces of weathering by rainfall and temperature. The maximum of rock breakdown, under clay saturation by nutrient cations and only small amounts of acidity, represent the fertile soils for protein production and good nutrition of higher forms of life.

in place of carbohydrate-producing ones for only fattening functions and less healthy animals.

The increasing temperature on going from the north to the south under the higher rainfall in the eastern states means that a different clay results from the development of the soil by that climatic combination. It is a clay which has little capacity to adsorb or hold acidity. Likewise, it can hold but little as its store of other positively charged nutrient elements when these are applied. Naturally, it has little of the original mineral reserves left to be decomposed. These clays themselves are decomposing more rapidly. Their soils, then, grow coniferous forests representing little protein potential even when cleared and not given heavier and more complete soil treatments regularly.

By superimposing the map of the precipitation-evaporation ratio over the one of the annual precipitation or rainfall, one can already see the soil map and understand why the mid-line running south from the northwestern corner of Minnesota to the southern tip of Texas divides those soils with calcium or lime (pedocals) to the west from those to the east (pedalfers), which are deficient in this element required in generous supplies for the soil to grow the legumes as high-protein forages by which the cow is most naturally fed. One can see on that soil map the limited soil types which, by their more complete stock of fertility, are the location growing the livestock by means of the proteins which only fertile soils can provide.

CHEMICAL TESTS OF THE SOIL TESTIFY FOR THE CLIMATIC SOIL PATTERN

Soil tests serve to emphasize the limited areas where the climatic forces are more nearly in balance, as they represent soil construction in our west and soil destruction in our east. Soil tests have cataloged the chemical properties of the soil, where the particular amount of rainfall is most effective, in giving that degree of soil development which grows the mineral-rich, protein-rich crops. Chemical examination of the supplies in the soils of calcium, magnesium, potassium and other positively charged elements, according to samples taken along an east-west line of constant temperature but increasing rainfall across Kansas and into Missouri, tell their story graphically in an interesting curve.

The increase of the rainfall up to 30 inches outlines a rising curve and an increasing soil's supply of active and exchangeable plant nutrients. This reaches the maximum at this rainfall figure in the chernozem soils of the northern midcontinent. But with higher rainfalls, the points on the curve drop to represent lower levels of fertility. This array of soil samples emphasizes the 25 to 30 inch rainfall as the maximum of loading the soil with the fertility for high-protein plants. Rainfalls higher than this value cause hydrogen, or acidity, to come into the soil to replace the plant nutrient elements by this non-nutrient one of similar positive electrical charge. Yet with the increase in rainfall there is an increase in the soil's saturation capacity. That increase in the soil's clay content means that more adsorption and exchange of nutrients are possible should we add such to the soil.

Thus, starting with no rainfall, its increase represents, first, an increasing soil construction by its making more clay and stocking that clay's adsorption-exchange capacity with active or exchangeable nutrient elements. That is the case up to about 25 to 30 inches of rainfall in the temperate zone. Then, second, with still higher rainfall there comes soil destruction. That means more water and its going down through the soil to carry nutrients out. It means more plant growth, more organic matter, more of it decaying, more acidity therefrom, and this, as hydrogen, replacing the calcium, magnesium, etc. in the soils to make them acid or deficient in the essentials for growing protein forages like the legumes.

All of this may well be theoretically and diagrammatically assembled as a rising curve and then a falling one in a bell-like shape representing the rise and fall of many of the soil properties, summed up in the general term soil productivity, so far as protein production is concerned. The curve for the soil's clay content takes a characteristic shape, resembling an elongated letter "S" tipped to the right at the top. With increasing moderate rainfalls, the soils have increasing amounts of clay. Then under the higher rainfalls, the clay is destroyed at about the rate at which it accumulates from rock decomposition. The curve for the total acidity starts at the 25 to 30 inch rainfall value. That is the value for the climatic factors at which the degree of soil development begins to be soil destruction. In moderate temperatures, the acidity curve is a parallel of the clay curve. For higher temperatures coupled with higher rainfalls, this climatic combination gives first a rise and then a fall of the curve for soil acidity.

This pattern of the effects of the climatic forces, in terms of different degrees of soil development, may be superimposed on the soil map by placing the vertical mid-line of the curve, which marks the 25 to 30 inches of annual rainfall, on the vertical mid-line of the soil map dividing the calcareous soils on the west from the non-calcareous or acid soils on the east. By such superimposition of the curve of chemical dynamics, brought about by the climatic forces as they are either soil construction or soil destruction in terms of potential for proteins in the natural vegetation in the various parts of the United States, we shall see the soil as the basis of many of the problems in livestock feeding for keeping the animals growing and healthy.

Additional study of the diagrammatic sketch helps us to see the soil processes of alkalization in the drier far west, and calcification (caliche) not so far west. The latter or dark brown soils are some of the pedocal soils on which

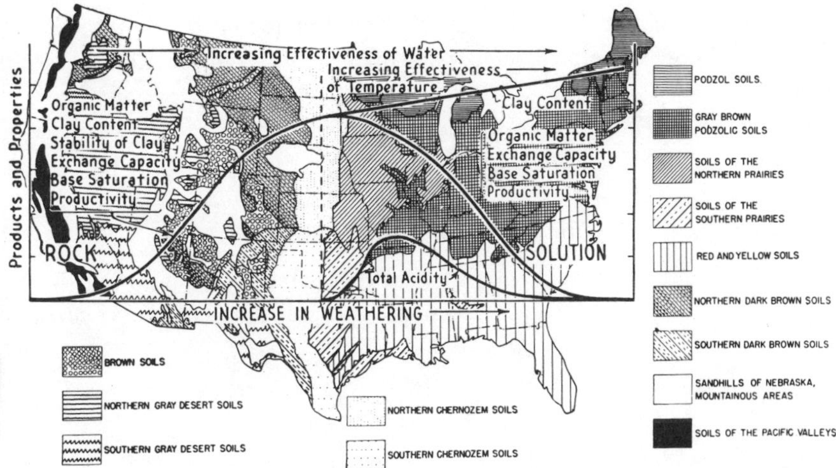

CLIMATIC AND VEGETATIONAL SOIL GROUPS OF THE UNITED STATES (After Marbut, 1935)

The midcontinent, with its concentration of farms and its high fertility, suggests the curve of soil development locating the higher production of protein in that area according to the chemo-dynamics of the soil.

the cattle range. Their scant herbage can be quickly over-grazed. It makes little annual growth because the weathering rate of the soil minerals, or the rate of the soil development under the climatic forces, is too low to build up much active fertility even "while the soil is resting." Also the deep, dark black soils (chernozems) at the eastern edge of the pedocals, once famous for their "hard" or high-protein wheat, do not hold up this protein output very long under cultivation either. In the last decade and a half the protein concentration in the wheat has fallen enough to close many of the flour mills. The millers report, "We don't get the quality in the wheat to make the quality of flour we once had." The "hard" wheat has become "soft" wheat, not because of change in plant variety, but because of the decline in the organic or the nitrogen part of the soil. The chances for the wheat to get its protein, like the chances for the cow to get hers, are going down. Such observations suggest that changes in the soil fertility may be classifying the western states in the same category with those east of the 98th meridian of longitude given to fattening, rather than growing, cattle as the more common practice with this class of livestock.

We are gradually realizing that the weathering processes in the soil, according to the climate forces of rainfall and temperature, come in to determine the quality of the crop production. Only as the rock is breaking down to make clay, and as that is completely stocked with all the inorganic essentials for the plant's creation of the complete proteins, can we expect to grow

The Climate Makes the Soil Pattern   17

healthy plants and healthy animals. It is the climate that makes the soil pattern and that operates the assembly line of agricultural production that feeds us all.

> *Though the Mills of God grind slowly,*
> *Yet they grind exceedingly small;*
> *Though with patience He stands waiting,*
> *With exactness grinds He all.*

Frederich von Logan. Sinngedichte.
(Longfellow, translated *Poetic Aphorisms;*
Retribution).

CHAPTER 3

Only Balanced Soil Fertility Grows Balanced Rations

IN STUDYING THE causal connection between the health of our animals and the fertility of the soil, via the crops the soil grows, we are apt to over emphasize the "ash" or the inorganic part of the crop coming from only the soil. But even that aspect, of itself, emphasizes the close connection. If any of the inorganic fertility elements are deficient or out of balance so that the plant is not truly healthy in any of its many functions promoted by those, then that soil shortage may produce the same condition in the animals confined to the deficient soil-plant area.

But animals must feed on organic substances. They take to bulky vegetation, such as grass, hay, etc., and not to salts and minerals, except as an act of desperation. Wildlife becomes marauders of crops on fertilized soil out in the open or in the absence of cover. They search out the soils offering better nutrition, even at the risk of their lives.

Animals need carbohydrates and fats for body energy; proteins for growth, protection and reproduction; inorganic elements as parts of their skeleton and parts of the organic feed substances; vitamins for help in using these; hormones; antibiotics; and possibly other higher complex molecular arrangements in organic compounds not yet established. By this latter fact and the necessity for the organic, we must connect animals with plants, naturally considered organic, rather than with the soil, so regularly considered inorganic and of the mineral kingdom.

But many of the organic feed compounds demanded by animals are the same as required by plants for their growth, protection, reproduction and survival. We give little thought to this fact. We emphasize the fact that plants compound their own carbohydrates from the elements carbon, hydrogen and oxygen (taken from air and water) by means of the enzyme chlorophyll. This accelerator of chemical reactions is itself a combination of both the inorganic and the organic. It is composed of an inorganic atom of magnesium (soil origin) around which protein-like and vitamin-like compounds are arranged. Yet we consider it an organic compound. The plant builds carbohydrates first as an anatomical structure (a) of roots going down into the soil to take the necessary inorganic elements from there; and (b) of its branches and leaves going up for carbon dioxide from the air and for sunshine, giving the energy required for photosynthesis. Some carbohydrates, *e.g.* sugars, are the starter compounds for their own conversion into proteins. Part of the carbohydrate supply is the energy material serving to bring that conversion about. This activity is a metabolic performance by the life processes within the plant. Protein production is, then, a *biosynthesis*, a synthesis by life itself, and not one brought about so directly by the sun's energy as is *photosynthesis*.

But even photosynthesis does not occur in the presence of protein. Plants are thus the compounders of carbon, hydrogen, oxygen and nitrogen into carbohydrates and proteins (part and parcel in all life processes) by the help of all the elements which the plants take as fertility from the soil. These contributions from the soil serve either as construction material or as tools in the plant's struggle to get its own proteins and energy in order to survive and reproduce its kind. In basic principles, the plant, like the animal, is spending its physiological activities in using thirteen known essential inorganic elements from the soil as the means of converting four from the air and water into the organic matter. Then by the fabrication and digestion of that, the plant grows, makes seed, and keeps the species surviving. Those same essential inorganic elements from the soil and via the plant are then in control also of the nutrition, and thereby of the health, of the animals fed by the plants.

THE ANATOMY OF THE RUMINANT CONNECTS IT VERY CLOSELY WITH THE SOIL

When we consider the cow, we note that her anatomical apparatus for digesting and handling bulky feeds connects her rather directly with the soil. As a consequence, our responsibility for giving her quality feeds is much lightened by her equipment for handling rather rough bulk. With the paunch as a converter, or fermenting vat, at the head-end rather than the tail-end of the cow's alimentary canal, the forage she ingests brings with it the soil, its microflora, and its fauna—all for preliminary convertive actions including both synthetic and digestive in the anaerobic conditions there. By that fact she lets us connect her and her health more closely with the soil. She profits by those symbiotic microbial relations. That is true, especially when her limited ingestion of urea, and its synthesis there to let it serve as partial protein supplement, bears the suggestion that its chemical structure carrying the amino nitrogen attached to the carbon serves so much more efficiently than does ammonia or other nitrogen forms not so similar to the amino acid structure of protein. That she has some advantages for survival on rough forages, through that special anatomical equipment and microbial partnership of the paunch, is suggested by the close companionship in which the pig and the chicken have always held her among the barnyard family, when they follow on her heels so closely to feed on her droppings of dung, but not of urine. When the pig and the chicken have their microbial, fermentive and digestive helps within the alimentary canal at the tail-end of it, they are not so closely connected with, nor so directly and completely supported by, the soil as the cow is. We may raise, then, the question, do they not have more diseases?

The distribution of the bison (also a ruminant) of earlier days over this country outlined a climatic pattern of health, then, for cows, according to the soil's development of its fertility growing the herbage as nutrition for them. With that as the basis for guidance, the feeding of cattle for their health does not necessarily require our being certain that we can prescribe the array of soil fertility elements complete enough, in list and function, to give certain specific results for economic management of herds of cattle. It would also be a vain presumption to believe that we have obtained that much organized knowledge about the soil and its capacity for creating healthy livestock. Rather, by studying the soils and the feed they grow naturally in the areas where, and by which, the bison was healthy in contrast to the fringes and areas beyond his concentration where he was extinguished, we can more nearly detect a few of the inorganic nutrients for possible soil treatments that may be limiting the forage quality for the health of our agricultural ruminant, the cow.

Even differences in the ash, or inorganic elements coming in the forage as cow feed from soils differing in their fertility make differences in its nutritional value. That those differences are detectable at the very first station in the digestive route, namely the paunch, was demonstrated recently. Higher amounts of volatile fatty acids resulted from the same region in the rumen (artificial) according as there were added the ash of alfalfa grown either (a) on the soil types shown by the crop quantity and quality to be more fertile or, (b) on the soils more carefully and completely fertilized.

From this prompt response in the early course of digestion to differences in the ash elements coming from the soil via the forages, there comes a strong theoretical suggestion that there should be corresponding differences in the inorganic elements in the cow's blood stream. If her paunch reflects the differences in soil fertility (either as elements or as the compounds created by them in the plants) by corresponding differences in the digestive and absorptive results from the bacterial flora active there, then presumably these differences should show themselves as balance or imbalance of the inorganic contents, and some organic, of her blood. This circulating liquid protein with its adsorption-exchange activities for the inorganic elements may tell us at this warm-blood terminal, or maximum construction of the food in body circuit, what the soil's adsorption-exchange activities contributed to the start of that construction or from that earthy point. Have we not neglected the possibility of using the chemical array of the inorganic elements in the cow-blood's composition as suggestions for soil fertility treatments to balance the cow's physiology, when such an approach was already found helpfully suggestive in connection with the problem of dwarfism? The cow's equipment suggests that because of her anatomical arrangement for digestion, outlining the creative biochemical transition from the cold, dead remains of the soil to the warm blood of the living cow, we can do much to make soil treatments reflect themselves in her nutrition, her health, and her procreation of her own species.

CARBOHYDRATES AND PROTEIN OR ONLY CARBOHYDRATES AS THE CLIMATIC SOIL PATTERN PROVIDES

The cow's wide adaptability to many kinds of roughages, because she is a ruminant, serves her well, provided she is permitted to range and select her grazing according to the fertility of the soil growing it over extensive land areas. On the contrary, those natural assets for offsetting deficiencies in her feed cannot come into play in her confinement by fence or stanchion. We need to see the forages she takes in terms of their balance of carbohydrates and protein (and all else) since those compounds within the crops fluctuate in their ratios, according to the variation in the ratios of the nutrient elements within the soil supporting the plant's synthesis of them.

We have long emphasized high calcium in the soil for growing more protein in the crops, especially legumes, on the humid soils of the eastern United States. We are now also connecting magnesium, phosphorus and nitrogen with the plant's processes delivering this organic body-building essential so commonly deficient in the ration. Then, too, we have emphasized potassium and even phosphorus as treatments for many soils for producing and mobilizing the carbohydrates. All this suggests the interconnections of the soil-borne elements in these two processes, making proteins and carbohydrates and the many other processes for growing nutritious grass or other good feed crops for the cow. More of the soil fertility, both in amount and variety, must be active when they grow mainly carbohydrates. It is a natural law that crop yields as bulk are not necessarily pushed higher when there is more protein in relation to the carbohydrates. There is not necessarily an increase in quantity because of an increase in quality of protein as nutrition.

According to the climatic development of the soil in the United States, the increasing rainfall (by traverse from west to east) up to about 30 inches gives increasing soil construction for growing protein-rich crops marking out the cattle-growing area. This is illustrated by the dark brown and chernozem soils in the United States. It is illustrated also in Canada, Argentina, South Africa, the U.S.S.R. and in other similar climatic settings by the soils on which beef production is the common use of the land. The drier areas, giving only a slight degree of soil development, are not the equal in this respect of the areas with a moderate degree of soil development near 30 inches of well-distributed annual rainfall represented by the chernozem soils in moderate temperatures. Under rainfalls higher than this figure, the high degree of soil development is soil destruction, so far as protein-producing crops can be the natural vegetation. Instead, there are crops of higher vegetative yields of higher carbohydrate but of lower concentrations of protein. Also there are the problems of the protein and "mineral" supplements to the normal forage crops if beef and cattle are to be grown, or even if to be fattened.

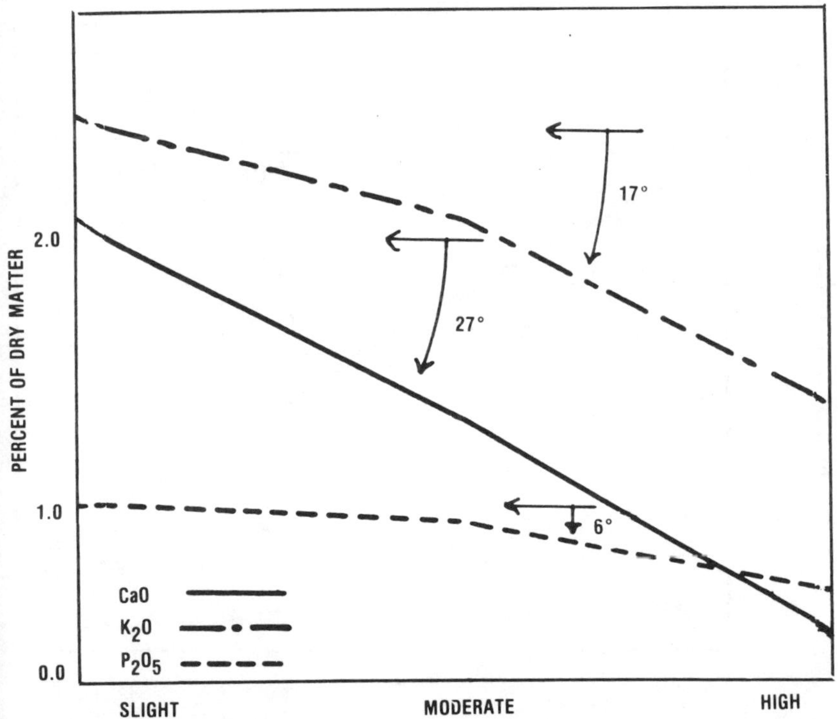

Calcium (CaO), potassium (K₂0), and phosphorus (P₂O₅), as percent of the dry matter in forages, all go lower as the degree of soil development under increasing rainfall goes from slight to moderate and to high. The calcium decreases more rapidly (27 degree angle) than either the potassium (17 degree angle) or the phosphorus (6 degree angle).

The problems of insufficient proteins and minerals in the cattle feed are regularly problems on the soils of slight and high degrees of soil development. These soils supply too little mineral fertility (inorganic or ash elements), or such too poorly balanced as a plant diet, to grow vegetative bulk creating enough proteins to balance the carbohydrates as a respectable cattle ration. Soils under either these low or high rainfalls represent problem areas in terms of fertility for growing naturally balanced rations for healthy animals. It is the soils of moderate degree of development which provide the protein-producing vegetation for healthy cattle. This is true the world over. Soils developed under rainfalls high enough and distributed regularly enough to keep crops growing throughout the season, have had some of the essential nutrients leached out (assuming they ever were there) to such a high degree that those remaining, are not a balanced diet for plant species converting enough of their carbohydrates into proteins to meet the feed requirements for

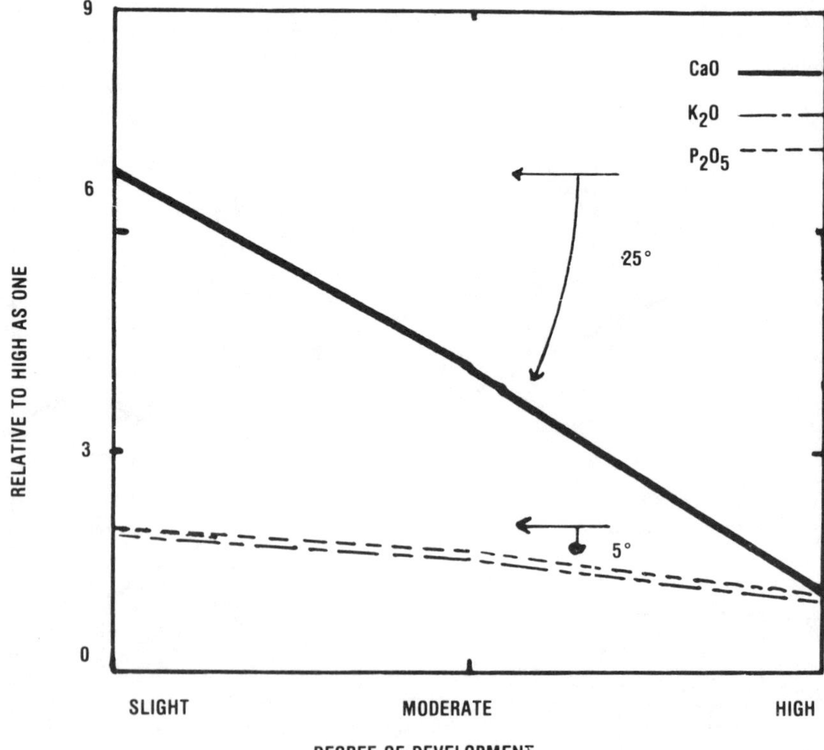

Calcium (CaO) is in wider ratio to the potassium (K₂0) and to the phosphorus (P₂O₅) within the forages as the soils are less highly developed. This gives more proteins, or a narrower carbohydrate-protein ratio in the forages as feeds.

growing healthy cows. Such are considered acid soils with increasing degrees of acidity (lower figures for pH) according as they are more highly developed under higher rainfalls.

CHEMICAL STUDIES OF THE FORAGES AND SOIL REINFORCE NATURE'S PATTERN FOR PROTEIN PRODUCTION

The chemical compositions of herbages taken commonly as feed by animals were studied according as they were grown (a) on soils of the west with but slight degree of soil development (38 species); (b) on soils of the midcontinents and a moderate degree (31 species); and (c) on soils of the east

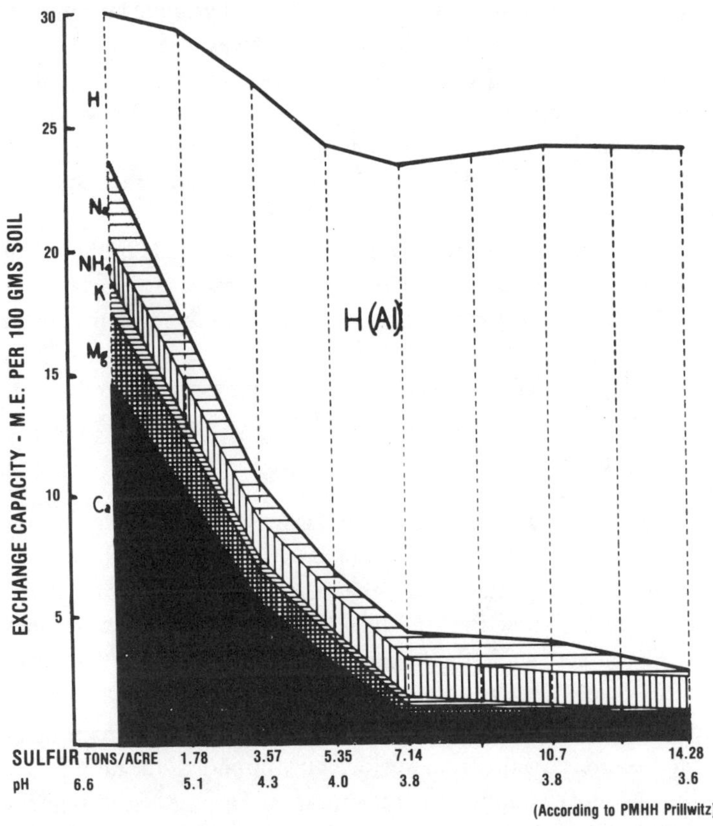

The exchangeable nutrient elements, calcium (CaO), magnesium (Mg), potassium (K), nitrogen (NH4), and sodium (Na), decrease rapidly while the nonnutrient, hydrogen (H, acidity), increases in soil given increasing degrees of development by more sulfur applied (increment = 1.785T/A.) The more rapid depletion of calcium than of the other nutrients emphasizes the narrowing calcium-potassium ratio as reason for the widening carbohydrate-protein ratio in the forages grown on corresponding soils.

and southeast illustrating a high degree of soil development (21 species). Their concentrations of calcium, associated with the protein production by these crops, decreased rapidly as percentages in the dry matter in following through the soil series in the above order of increasing degree of soil development. Their concentrations of potassium, associated with the carbohydrate production by the crops, decreased also but at a much lower rate than that of the calcium. There was then a wide calcium-potassium ratio in the composition of forages grown on soils of slight degree of development. This ratio became narrower as the degree of soil development went higher and higher with the increasing rainfall.

The data from these studies are assembled with graphic representation. In going from *slight* to *moderate*, to *high* in degrees of soil development, the percentages of calcium oxide in the dry forage shown in the graph give the following values: 1.92, 1.17 and .28 respectively or in an almost straight line. The percentages of potassium oxide on its graph were 2.44, 2.08 and 1.27 respectively. Those for phosphorus oxide decreased at a rate much like that for potassium with the values .78, .69 and .42 respectively.

The graph or line for the decrease of the calcium oxide in the forage feeds on going to the higher rainfalls, which represent more crop yields as tonnage but less of this element per mouthful for bone-building and for all the other functions in health connected with lime, goes down at an angle of 27 degrees from the horizontal. The angle of the curve for potassium decrease with more development of the soil under more rainfall is 17 degrees. The angle of decline for phosphorus in the forages under the same climatic shift is but 6 degrees from the horizontal.

If we should set the concentrations of each of these three soil borne essential elements in the forages at unity, or one, for their growths on the soil of high degree of development, then the other values in relation thereto (see diagram, page 23) show that the line for the calcium oxide declines at an angle of 25 degrees, while the angles of decline for both the potassium and phosphorus oxides are the same and at the value of but 5 degrees.

Here, then, in these shifts in degrees of development of the soils growing the plants, there is a shift from a wide calcium-potassium ratio to a narrow one in the chemical composition of the forages. There is also a duplicate shift in the calcium-phosphorus ratio, (a) as there is more yield of bulk in general; (b) as there is a decrease in the protein concentration and an increase in the carbohydrate of the feed; and (c) as there is a higher degree of the soil development. Thus the more rainfall, to which more abundant tonnage yield is ascribed, is making more carbohydrates. This is readily piled up by photosynthesis using mainly air, water and sunshine. But that higher rainfall is making less protein because it has given depleted soil fertility (more soil acidity) determining the crop composition via the soil and not via the rainfall alone.

That the soil under the increasing rainfall represents similar downward shifts in its active or exchangeable nutrients with higher rainfall, and thereby higher degrees of soil development, has been demonstrated by using increasing amounts of sulfur; oxidizing in the soil to give more sulfuric acid as one might use increasing amounts of organic matter oxidizing there to give increasing amounts of carbonic acid. As the soil had more sulfur to make it more highly developed (shown by the decreasing pH values in the figure on page 25), the calcium was leached out under constant rainfall much more rapidly than was the potassium. Here, then, the narrowing ratio of calcium to potassium and to others of the exchangeable nutrients in the soil is the reason for the corresponding shifts in the ratios of those same elements in the forages. The degree of the soil development sets the fertility pattern of elements either in balance, or out of balance, for crop composition of proteins as well as carbohydrates. The forages then set the pattern of their composition according to the pattern as the soil fertility demands it.

Consequently, as either nature develops the soil more highly and thereby depletes the fertility more, or as we cultivate the soil and remove more of it, there is less of the proteins produced by the forage. That may even give larger tonnage yields of it with certain higher degrees of soil depletion. Therein lies the dangerous deception for animal health in looking to soil areas of much rain and much crop. Where rainfall has always been enough to produce much forage, there the virgin soil was probably growing forests; or for large forage yields such soils usually mean protein deficiencies as feed for healthy cows. Such soils grow good yields of carbohydrate crops for fattening the older animals, provided they are grown with good health and its physiological reserve on some other soils. That simple fact promotes the movement of cattle eastward from the west in the United States (also in Argentina) for the fattening of them. That is also a movement to the markets. In this latter movement, the economic thinking smothers out the great biological fact that the movement to the soils, serving mainly in fattening for the animal, is moving it, and the species as a whole, to a lower level of health.

The constant aim at higher crop yields per acre, and the introduction of exotic plant species solely for that purpose, permit the decline in both soil fertility and nutritional quality of the crops to go on without our concern about them. It is because of that confusion that we have not recognized the great ecological fact that animals, in their nutrition, their health, and their reproduction, depend on the higher fertility levels of the soils according as the climates have built such in proper balance.

HALF-LIFE OF SOILS

I have excellent data on the half-life of our soils. You see the soil is like a radioactive element newly created. When this soil was balanced out in man's absence, and before man took it over, it was virgin soil. It was in equilibrium with the forces of soil development and leaching. If you start with the desert in the west, on the east side of the coast ranges—because water has all been precipitated on the west side—that's the raw rock with a slight weathering. As you come east, then it is heavier rainfall, and you develop the soil into more than a desert. And the American bison lived where conditions were about balanced, and that's a little above 25 inches of rainfall. Because when you go above 25 inches of rainfall, you began leaching. But at 25 inches, you're just about balanced. That buffalo was smart. He had mineral-rich soil and not mineral-leached soil.

And it's been grown with crops that suck only the back teat, we'll say, and remove certain elements more than others. The buffalo didn't go far east and west, but north and south. He went with the winter and summer, back and forth. He went long distances north and south, but he didn't migrate far east and west, because he would have gone to less rainfall, and more rainfall.—*Dr. William A. Albrecht, in an interview with Acres U.S.A.*

CHAPTER 4

Nutritious Feeds via Soil Fertility and Not Plant Pedigrees

WE GROW CROPS to provide the organic substances in hays, grains, forages, etc., which can be nutrition for our farm animals. Feeds are photosynthetic products containing carbohydrates, fats, proteins, vitamins, etc., but less than 10% of inorganic substances (ash, minerals) originating in the soil. Crops are the creative connection between the soil, which supports all terrestrial life by those contributions of fertility, and thereby all included in the term animal health. Consequently, any kind of life will be either healthy or unhealthy according as the fertility of the soil determines.

Animals are organic bodies which feed on the plants because these are organic substances offering energy and growth values for nutrition. Plants, however, must feed mainly on inorganic elements which determine the photosynthetic and biosynthetic plant processes compounding the crops into organic substances, not necessarily nutritive ones. When, in its struggle for nourishment, the animal is compelled to bypass the plant's faulty delivery of the animal's inorganic requirements and to eat the necessary inorganics directly from the mineral box, *i.e.*, those outside of the plant compounds usually providing them, we may well consider the animal behavior an act of desperation because of defective health. Animals are struggling to be healthy. Apparently they aim to be so via their normal nutrition. They know nothing of crop pedigrees or plant variety names.

AS PLANT YIELDS GO HIGHER, PROTEINS GO LOWER

Domestication hinders the animal in its struggle to be healthy. While we have many crop plants, we have emphasized and adopted each particular kind mainly for its production of more yield as bushels or tons per acre. The introduction of new crops, and the importation of foreign plants, has been guided all too often by the single creation, namely, "Will it give more yield!" We have not recorded the plant's many physiological activities related to the soil fertility as exhibited in the natural setting of the plant in question. We have not observed closely whether the plant was making its attractive exhibition on a soil developed to a lesser degree under moderate climatic forces, or on one developed to a higher degree by high rainfall and temperature. We have, therefore, not distinguished between the mineral-rich, protein-rich crops of lesser bulk, common in the former setting, and the protein-poor, mineral-poor but highly carbonaceous ones of greater bulk so common in the latter. Nor have we realized and accepted the simple fact that for the nutrition of the cow, the former kind of plants, grown where the climatic conditions developed the soil to produce less quantity but more quality, represented better nutrition and thereby better health for the cow, while the latter kind of plants with higher yields but lower concentrations of proteins—and of much else which cannot be synthesized on such highly weathered soils—should suggest deficiencies in animal nutrition and the consequent troubles in animal health.

As we follow the larger pattern of different crops across the United States from the midwest, with its low rainfall, and go to the east, with its high annual precipitation, we have increasing yields of vegetative bulk per acre. At the same time, there are decreasing concentrations of protein and inorganic elements in that vegetation. The growth is simply more carbonaceous, or more fibrous and woody. In our feeding of it we are constantly disturbed by the needs for protein supplements and extra "minerals." That need shows up even when only fattening an older castrated male. Problems of difficult calving, and defectives at birth to include not only weaklings but dwarfs and other biological irregularities, are occurring at an all too high a percentage in the animal population on those soils developed more highly under heavier rainfalls, even at moderate temperatures.

Also, under the higher rainfalls as we go south in the eastern United States to higher temperature as well, the virgin plants themselves are those which reproduce vegetatively more commonly. This fact indicates their failing struggle to make enough protein for reproduction by seed. Virgin vegetation there suggests its failing health when we observe the parasitic attacks from Spanish Moss on the lower, more mature branches of the younger oak trees. Even the southern pine is taken by the Spanish Moss. Plants are not well-fed and healthy on those highly developed soils. Poor plant health is indicated by poor reproduction for survival of the species.

Yet in those areas of higher moisture and temperatures, we find tremendous growths and underground yields of such carbonaceous producers like the sweet potato root. We find also big tonnages above ground of saccharine grass vegetation like the sugarcane. When considered for their bulk per acre, we can understand why such crops are commonly characterized for the ease with which they can be grown. They are also commonly marked out for their low values as feeds, unless highly supplemented by both the proteins and the minerals, and then when used mainly for our bulk handling ruminant, the cow.

THE LOW QUALITY AS NUTRITION AND HIGH YIELD OF BULK DEMONSTRATE THEIR MATHEMATICALLY CLOSE RELATION

The different crops have fitted themselves into a precise pattern in nature according to their requirements for soil fertility. This arrangement has each species located, for its special advantage, on the soil developed by the climatic forces to the particular degree better suited for it than for any other. Plants are located in that natural geographic pattern not as the pedigree or

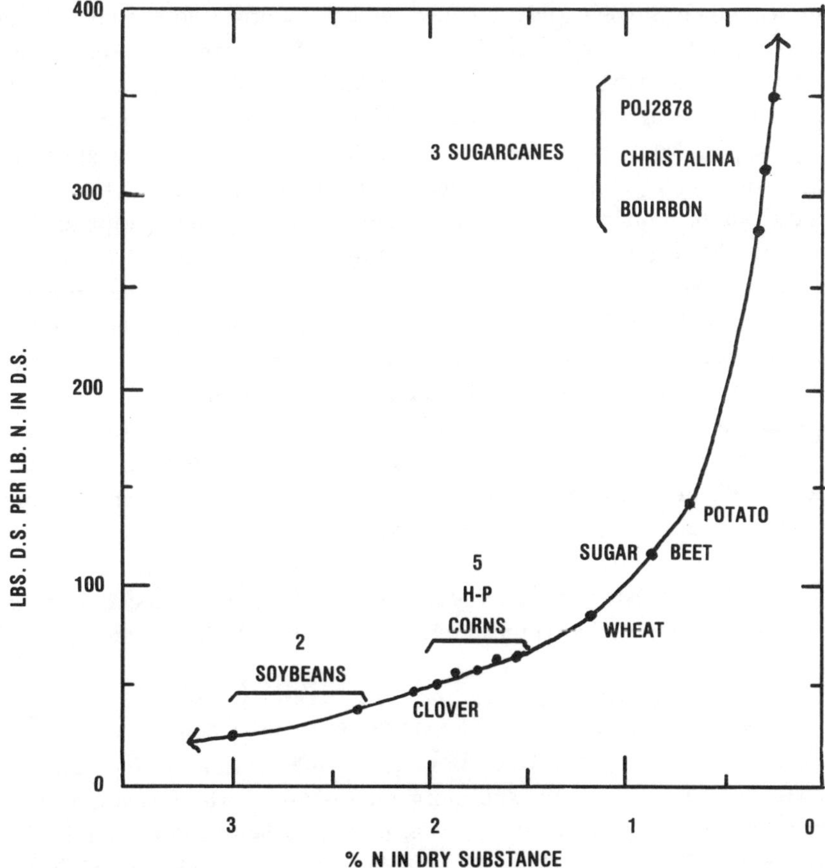

The yields of dry plant substance by some common crops as a function of the concentration of the nitrogen (protein) as percent in the dry substance.

the plant name designates, but according as the soil fertility serves to nourish them for their survival via their own growth process. The study of this ecological pattern offers much for our better understanding of some agricultural problems should we fit crops and animals more nearly to the soil offerings.

That the increasing yield per acre (dry matter) of different plant species in their ecological arrangement should mean that it is made up of more carbohydrates but of less and less protein, or less and less nitrogen (N), has been worked out to a mathematical refinement in the so-called *Inverse Yield—Nitrogen Law of Nature*, by O.W. Wilcox, published as a mimeographed tract in June 1956. This arrangement shows how the tremendous yield of bulk per acre and the very low concentration of protein, (N times 6.25) in the case of sugarcane as one of the grasses, fits the curve of this law as do the high concentrations of protein in the legumes, like soybeans, but with their smaller yields of hay per acre. Corn, another one of the grasses, can have considerable concentrations of nitrogen. However, the introduction of its hybrids has reduced that while the starch and fodder yields have gone up. Hybridization has been the equivalent of pushing the physiological performances by the corn plant down to make it duplicate more nearly those of sugarcane. By this manipulation, we have pushed this crop's production of protein nearly down and out for growing young animals.

As another legume, the clover, for example, is more nearly like the soybean. They both take some nitrogen from the air to supplement that from the soil. Thereby the legumes have an advantage over the grasses, not so equipped physiologically, to draw on the seventy million pounds in the air over every acre. Other crops, both legumes and non-legumes, fit themselves naturally into this orderly mathematical arrangement. This serves to show clearly that some plants pile up the carbohydrate bulk in their vegetation readily with little requirement from the soil. But plants which make more proteins per acre require much of a wide array of soil fertility. Also, those consume much of their own carbohydrates, partly as the chemical starting compound and partly as the energy source by which the life processes of the plant bring about the biosynthesis of converting carbohydrates into protein.

THE COW MUST EXERCISE HER CHEMICAL SENSE
WHEN SHE CANNOT HANDLE OUR CROPS OF BULK UNLIMITED

As feed, our crops differ widely in what they offer as guarantee for animal growth, that is, the proteins and inorganic essentials; while most any crop carries more nearly constant amounts of the carbohydrates, that is, its starches and fibers for filling and fattening. It is in our viewing of feeds, mainly for energy or for putting on fat, that we are bringing on the danger to the animal's health. Good health is not guaranteed by the calories of the feed. These are the values which are pushed up when we make hays from the more mature plants in order to get larger tonnages per acre and then find that we have pushed its nutritional values for growth and health down to where we must call in more protein and mineral supplements. The Swiss dairyman, who cannot import such and grows little or none in crops other than grass, must cut this in an immature stage for hay. He cuts it when its protein concentration relative to its carbohydrate is high. As a result, his hay is feed, not just filler. Young plants deliver more growth values per mouthful for the cow. Nature suggests this for us to notice when, by evolution, the parturition for both wild and domestic herbivorous animals comes naturally in the spring season to guarantee the young their required proteins for growth.

This fact is well shown by the changing chemical composition of the plant, taken in its entirety, at different periods in its growth cycle. Alfalfa plants, for example, harvested at a height of six and a half inches, contained 28.0% of protein and 12.25% of ash or minerals. But when taken at full bloom, the corresponding values of these essentials for growth dropped to 12.99 and 8.04%, respectively, in the dry substance. At the same time, the woodiness in this potentially proteinaceous forage, if taken at the proper harvest date, rose in its fiber concentration from 12.35 to 36.2%.

Thus by changes in composition with maturity of the growing plant, the increasing yield would call for trebling the amount consumed by the animal to offset the decreased feed value from extra fiber content. In terms of decreased protein concentration, the amount taken would need to be more than doubled. Even in terms of lowered concentration of ash with maturity, the ratio would need to be increased by 50% in all these cases to maintain the

---

Chemical Composition of Common Forages

	Percent-Dry Substance							*PPM-Dry Substance						No. of samples
	Ash	Ca**	Mg	K	P	S	Cl	Fe	Mn	Cu	Co	Zn	B	
Timothy	5.93	.32	.17	1.65	.15	.19	.32	110	75	10.5	.07	17	5	20
Red clover	6.30	1.26	.36	1.40	.17	.14	.17	129	38	10.5	.15	17	12	28
Alfalfa	7.60	1.31	.29	1.79	.23	.26	.24	181	29	8.9	.10	17	13	66

Missouri Agricultural Experimental Station, Reserve Bulletin 533, 1953.
**Ca—calcium; Mg—magnesium; K—potassium; P—phosphorus; S—sulfur; Cl—chlorine; Fe—iron; Mn—manganese; Cu—copper; Co—cobalt; Zn—zinc; B—boron.*

---

equivalent of the initial nutritional quality per mouthful. In terms of the cow, unfortunately she cannot treble her consumption capacity to make up for the low nutritional value in the larger hay harvest per acre desired by her owner, but not by herself. She cannot be expected to be a hay handling machine only. While he emphasizes the yield and pedigree of the crop, she is scrutinizing it for values which guarantee her good nutrition, good health, and potential offspring.

As an illustration of the differences between crops in their composition of the chemical essentials drawn from the soil, we may well look at some Missouri data for timothy, red clover, and alfalfa (see the table on page 33).

These are three common hay crops. One is a non-legume and two are legumes. Their differences in contents of the essential elements, both major and trace, may be more significant than the small figures in the absolute represent.

We may well note that timothy has five times as much potassium given to carbohydrate processes in plants as it has calcium which is connected with protein production there. Calcium is also connected with the mobilization into the roots of many of the fertility elements contributing to that plant process. In the case of the two legumes, red clover and alfalfa, there is no such wide ratio. Rather, the calcium and potassium in the clover are more nearly the same. This points to the high lime needs (calcium) by legume crops for their high capacities in protein production.

But calcium alone is not enough for protein output by legumes. Too long have we believed that in the eastern United States and the highly developed or so-called acid soils, one needs to add only lime to grow legumes. These crops are high in the other lime-essential, namely the magnesium-lime element. The higher requirements for magnesium by soybeans are pronounced. Clover and alfalfa are also usually higher in both sulfur and iron than is the non-legume timothy. The legumes are usually lower in chlorine. They also have wide calcium-phosphorus ratios, usually 8 to 1 for red clover and 5.4 to 1 for alfalfa. For timothy, this value is but 2.0 to 1.0.

If the cow's blood has a calcium-phosphorus ratio near that of the human blood, namely 2.0 to 1.0 or 2.5 to 1.0, we can appreciate her physiological struggle to balance these two essentials in her blood stream when their ratios differ so widely in the crops from which she must supply those elements to the blood. When we confine her for months at a time to a single crop of too wide or too narrow a ratio in these respects, should we be surprised that she breaks down in her health? Can she overcome our pronounced concern about plant pedigrees and tonnage yields when her health is concerned with plant composition from fertile soils growing a balanced diet for her?

Chemical analyses show that the minor trace elements deserve attention too since they are connected with vital processes of both plants and animals. Cobalt has taken on much attention since we have learned that it is needed for animals. Only recently has it been shown to be essential for the nitrogen-fixing algae (*Anabeana cylindrica*) in its color development but not necessarily for the growth of this single-celled plant. Here we come to the importance of the fertility elements in the processes of life not connected with increases in yield or bulk. Another trace element, chlorine, has shown itself essential for plants. It has long been such for animals. It has been supplied them as a supplementary "mineral" in the form of common salt. Cobalt is higher in red clover and alfalfa than in timothy. The same is true for boron. This element has not yet been shown necessary for animals, but is needed by plants, especially those using atmospheric nitrogen and making more protein of themselves. In the case of manganese, the timothy has a higher concentration of this essential trace element than have the red clover and alfalfa. This is expectable for timothy grown on the more acid soils where the manganese is commonly more active.

These trace elements are gradually demonstrating their differences in concentrations in crops too, as we note that they may not show up through their efforts on yield increase, but rather through some of the life processes or the physiology. It is also by way of the life processes of the cow that the trace elements show their efforts more than in gain in weight. The cow's discrimination between forages is according to quality as nutrition and not according to quantity grown per acre. She seems to exercise a more discriminating chemical sense of nutritional quality than we in our chemical considerations of crops yet appreciate.

THE BISON ALSO NEEDED TO DISCRIMINATE BETWEEN FORAGES FOR QUALITY RATHER THAN QUANTITY

Virgin vegetation over even a limited area on the range of the west may show wide variation in chemical composition. A recently presented study of the forage containing 15 native plants of Idaho showed the concentrations of calcium in the virgin non-legumes ranging from .50 to 2.84% of the dry matter. For the phosphorus, the range was from .13 to .38. Taking the low figure as one, then the range for calcium goes from that to 5, while for phosphorus the range is from 1 to 3. Viewed similarly in case of trace elements, cobalt varied from 1 to 4.5; copper to 4.0; manganese to 7.4; and zinc to 2.6. Those less-weathered soils of the west grown to alfalfa give this crop calcium concentrations of 2.40, 2.28, and 2.40%, as sets of averages, and phosphorus concentrations of .42, .54, and .55%. These are roughly twice as high as the corresponding values for this legume feed grown on soils of Missouri.

Given average figures for chemical analyses of forages, one dare not forget the wide variation in the inorganic composition of a single crop because of the difference in the fertility of the soil growing it. The plant is confined to a limited area of soil. It cannot break out and feed on the neighbor's land. It is, therefore, the soil and not the plant pedigree that decides what nutritional values and chemical composition the crop has.

Though the American bison brought fame to his "buffalo grass," even that reputation does not guarantee constancy in chemical composition of this grass (*Buchloe dactyloides*). This fact was established by research of the Soil Conservation Service, United States Department of Agriculture, in their analyses of samples collected in close proximity in the western Gulf Region. Those samples showed a range in protein from 1.5 to 16.0%; in phosphorus from .03 to .31%; in calcium from .07 to 1.58%, and in potassium from .10 to 2.17% of the dry matter. These higher values were as much as ten times the lower ones in the cases of the protein and the phosphorus; 20 times for the calcium; and 21 times in the case of the potassium.

With such wide differences in the protein as an organic essential and even in the ash with its "mineral essentials," would the bison of the past or even the cows of today have survived had they been confined to these lower values, or had they not been distinguished by their selections between such extensive variations in plant chemical composition?

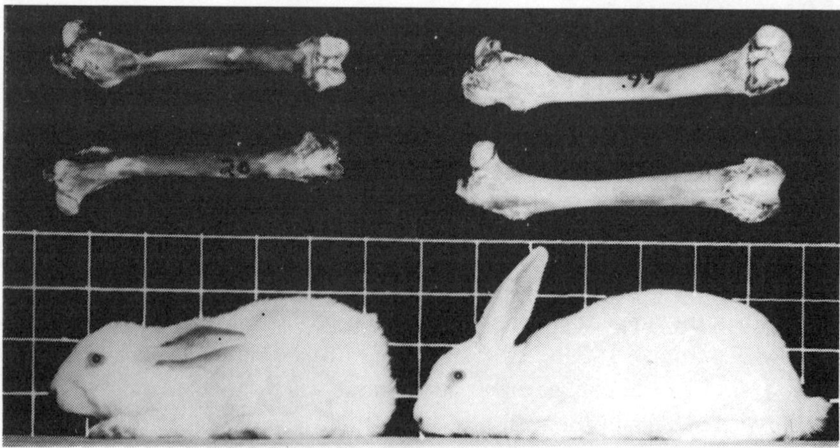

The weanling rabbits had the same pedigree, so did the crop plants making up the hay, but treatment of the soil with some extra fertility to grow better feed made the rabbit on the right different in appearance and body structure as the bones also illustrate.

Nature's pattern of soils developed from rocks by the climatic forces outlines the pattern of crops, offering those of much carbohydrate growth and little more than fattening power for livestock in the larger yields of either bulky vegetation or starchy grains. Crop yields of such chemical composition tell us that the plants are struggling for proteins to make germs to keep the species reproducing. Since crops have been selected and propagated for big yields under many kinds of protection, they have become less healthy. In the virgin condition, only plants making protein for protecting themselves against diseases were surviving, and then only on soils offering enough dynamic fertility to make the biological protection possible.

Nature's soil pattern and corresponding crop pattern demonstrate the simple fact that rainfall enough to water crops regularly and to give big annual yields means soils highly weathered and growing vegetation of low protein and low inorganic contents, or poor feed for growing and reproducing herbivorous livestock. If the virgin wildlife crop did not give evidence of the high protein value in the forage crops because of   fertile soil-feeding grass-eating animals, shall we hope to be successful by merely transplanting the cow? Must we not also modify the fertility of the soil so that it can grow not only enough protein but also is complete in quality? Can we expect the cow to protect her health by feeding only on bulky carbohydrates as feed from transplanted and exotic crops which could set up only a woody plant structure but make no seed?

Transplanting a grass agriculture from the bison area, according to nature's soil pattern, may offer hope for vegetation cover against erosion of a highly developed and worn-out soil mechanically reclaimed against that destructive process. But the simple fact that a crop grows is no proof that it is feed for a cow. Nor is it evidence that it will help protect herself against disease, even if her struggle for that is supplemented in that problem area by technological helps like the hypodermic needle.

The cow is not a mowing machine nor a hay baler. She cannot tolerate starvation on a paunch filled with no more than big yields of bulk. She is a set of complicated biological processes using feeds to do more than fill, fatten and gain weight. According to her role in nature's pattern of the many life forms dependent on the soil for creation and survival, she is contributing to the life stream of domestic livestock, which seems to have continued its flow of the past in spite of us, rather than because of us, as soil managers for crops creating nutrition rather than just growing.

Under starvation of our crops for protein on highly weathered and exploitively cropped soils, shall we not expect starvation and deficiencies by our animals? Under their stresses and struggles to survive, we may well look to faulty and failing reproduction as the first symptom of those deficiencies in nutrition in case of both plants and animals. There is no escape by ascribing the trouble to the plant's or animal's pedigree, or to their line of breeding. The spermatozoa, the ova, the chromosomes, and the genes are all highly specific proteins. The gene, therefore, may suffer deficiencies too. Such are losses of transmissible characters via losses of protein characters. Yet the gene, too, struggles to keep the stream of its own life flowing, which may mean accumulated losses all originating via nutrition as feed and thereby via the soil fertility. The pedigree of the plant does not guarantee the quality of the crop as feed for our animals. Only a fertile soil does that.

CHAPTER 5

Calcium, the Premier of the Soil's Nutrient Elements

THE SELECTION OF crops for big yields of bulk has reacted to take out of the stream of vegetative life much (a) of the power of plants for protein production, (b) of their protection against invasion by microbes, viruses, and other foreign proteins or what we call "diseases" and (c) of their capacity for fecund reproduction, or the guarantee for the survival of the species. These several capacities for plant species survival rest in the tissue proteins. Energy food compounds alone, like both carbohydrates and fats, do not (a) support cell growth, (b) give immunity to microbial or virus afflictions or (c) provide fecundity in delivering offspring. Protein-rich crops require soils high in calcium, the premier among the soil-borne inorganic essentials for the life of animals and man. Animal health can scarcely be expected unless the soil and what it produces are well stocked with calcium. Even this must be in balance with all other nutrients.

We have long been liming the soil to grow legumes. But we have erroneously valued the limestone, *i.e.*, the calcium carbonate, in that service because the carbonate part, or the chemical anion, was a means of reducing the concentration of the hydrogen ion, which is the acidity element of the soil. We failed to see the calcium, the chemical cation, as the foremost nutrient element required as an addition to the humid, highly-developed soils if they are to grow protein-rich forages. Readily manipulatable laboratory gadgets, which measure the changes in degree of soil acidity (pH) to a find point in connection with limestone applications, kept up the emphasis on limestone to correct soil acidity and therefore to grow more legumes.

The increasing failures of legume growth correlated with naturally increasing degrees of soil acidity in the ecology, or natural pattern of crops, were erroneously taken as proof that the *presence* of soil acidity, namely ionized and active hydrogen, was the *cause* of the poor growth of the protein-producing legumes. It was only after diligent study of calcium as nutrition for plants fixing nitrogen and synthesizing proteins that the calcium of limestone moved from its low classification of a "soil conditioner" and a "reducer of the hydrogen ion of the soil" to become the prime minister in plant nutrition. Calcium is more than a cheap carbonate as ammunition in a fight to neutralize the soil acidity.

This classification as premier is fitting for several reasons. Calcium is required in the soil's offerings, or its suite, of exchangeable cations, *i.e.*, ions of positive electric charges to a higher degree of saturation of the soil's holding and exchanging capacity than any other chemical element. It is the index element of the soil's degree of development under the climatic forces. Thereby it is the major criterion in the classification of soils on their fertility basis for protein production. Calcium is the major "ash" element in the highly proteinaceous bodies like those of animals and man. It mobilizes other ions, those both positively and negatively charged, into the plant by its service in the soil and in the plant roots. It is prominent in soils of which the physical condition as well as the chemical is favorable for excellent crop production. Also, calcium is the prime element in the nutrition of the plants which are giving us the various organic products serving to bring good animal health via good animal nutrition.

Increasing the calcium gave increasing protection to soybean plants against fungus attack. Giving the sand more hydrogen-calcium clay of pH 4.4 (left to right, visible in bottom of glass containers) gave better plant nutrition and better health. For these plants, "To be well fed is to be healthy."

Living on the soils of the humid region, our animals, like ourselves, do not get enough calcium regularly to prevent nutritional deficiencies. Dr. Babcock, et al., has reported that for humans, even the well-nourished, well-developed males are apt to be deficient in thiamine, riboflavin, ascorbic acid, calcium, and phosphorus. Carbohydrate crops and highly developed soils do not deliver enough calcium regularly for the growing animal body. The mineral box is a poor substitution of the inorganic calcium for those many organic combinations into which growing plants compound it. Legumes are a good illustration of excellent suppliers of calcium in those organic combinations for animal use.

The significance of calcium in those plants creating protein to high concentrations has never been pinpointed to any particular physiological function by this element. It does not appear in the final protein product of either plants or animals. It serves in the case of some non-leguminous plants to remove some of the metabolic byproducts by combining with them to make them highly insoluble. Thereby, however, it takes itself out of digestive service in the animal body. This is well illustrated by some of the vegetable greens of the goosefoot family (New Zealand spinach, Swiss chard, beet greens and spinach), which produce much oxalic acid to make their contents of calcium and magnesium indigestible as oxalates. The greens of the mustard family (kale, mustard, and turnip tops) do not produce the excessive oxalate as probable metabolic byproducts, hence their calcium and magnesium are digestible. Measuring the calcium by ash analysis may be a deceptive measure of its nutritional value in the crop. There are many plants which produce oxalic acid. They are grouped under the oxalis, or the woodsorrel family, which may contain their calcium in that form inactivated by combination into the oxalates. Of itself, calcium is not so highly active chemically, hence is considered an alkaline earth rather than an alkali.

Because of some of the chemical properties of calcium, the soils growing better legumes must be stocked with exchangeable calcium on the clay colloid complex to roughly 75% of its capacity. This large amount of potentially active calcium, for exchange to the root for the hydrogen this plant part offers in trade, means that calcium dominates in the soil's supply or the suite of exchangeable and available elements.

Magnesium, coming next in that category, needs to fill no more than 10% of the exchange capacity. It is only about 1/7th or 1/8th as much as the calcium in the suggested soil's balanced ration for legumes. Then potassium, coming third, drops to 3 to 5%. It need only be between 1/25th and 1/15th as much as the calcium, with the trace elements and some unknowns making up the balance on the clay colloid as required plant nutrition. In a soil carrying a balanced ration for growing legumes, little of the soil's capacity is left to be occupied by hydrogen (a non-nutrient from the soil) which comes in to replace these other cations as the soils become highly developed or more acid and less productive. In terms, then, of the amount needed in the soil in active form, calcium stands at the head of the list. It is the premier of the cationic crowd of elements for the nutrition of healthy plants and animals.

In terms of the amounts in the animal body, calcium is also at the top of the list of essential ash elements coming from the soil. It is located mainly in the animal skeleton. But it is functioning there in more ways than just as so much skeletal reinforcing in a mass of soft tissue. Calcium is highly mobile into, and out of, that bone reserve of itself in combination with phosphate. That mobility is unique. Since calcium is divalent and not so soluble, while phosphorus is trivalent and also combined into many even less soluble compounds, these two take hydrogen, or acidity, into their combination, but thereby bring about more solubility and higher activities of themselves. This

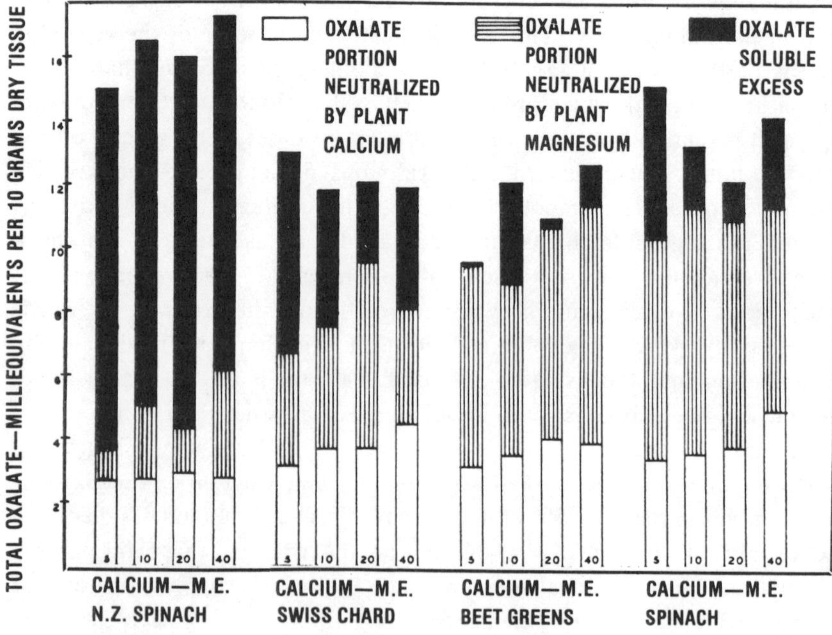

The probable disposition of oxalate in some vegetable "greens," grown on soils with variable tests of calcium, indicates the excess of oxalate (black sections) beyond that needed to make the calcium (white section) and magnesium (ruled section) insoluble and indigestible. It indicates the deception in measuring ash calcium for its nutritional value.

results in a series of combinations of calcium and phosphorus ranging from the insolubles of the bone to the partly acidic but more active combinations by which some processes are buffered or protected against sudden chemical shifts in acidities and solubilities which would result in shock.

Milk fever suggests that the shifts of stored calcium into the larger, active supply for milk production after calving are too slow or are disrupted, or that this element is involved in disturbed body processes about which we know all too little to prevent this trouble. In this situation, which follows usually so close to parturition though possibly at other times, we appreciate the premier role of calcium (supplemented by other nutrient elements) especially when the organic combination of it as a gluconate does so much so quickly in literally resurrecting a cow, or in snatching her from death by milk fever.

CALCIUM EFFICIENCY IN PLANT AND ANIMAL NUTRITION IS IMPROVED BY THE PRESENCE OF SOME ACIDITY OR ACTIVE HYDROGEN

It is significant to note that both calcium and phosphorus in chemical combination with different degrees of acidity, hydrogen, are essential compounds in the animal body. It is a fact that some acidity or hydrogen in the soil is important also in the nutrition and physiology of the plants, especially legumes. When hydrogen is completely removed from the soil and when a surplus of calcium carbonate, or limestone, is maintained to prohibit any acidity, then there prevails a condition highly disturbing to the nutrition of better forages. Liming the soil is not a case of "Where a little is good, more will be better." More may be more damaging to animal health via the crops than we yet recognize as damage to them. Some cases of dwarfism in cattle have suggested their relation to such soil conditions.

Experimental studies of growing crops by using carefully controlled clays for nourishing them have shown that the legumes must have that clay's exchange capacity saturated highly (75 to 85%) by calcium to have them grow by fixing their own nitrogen from the atmosphere. By similar experiments it was shown that when some acidity, hydrogen, is also present or accompanies this large supply of calcium, this premier element is moved into the plant to a higher degree, or more efficiently, from the soil's available supply. This efficiency is higher than when the calcium is accompanied by other elements (barium for instance) to the degree of excluding the acidity, or hydrogen, completely. Therefore, some acidity in the soil is good company for the calcium if this and other nutrients are to be taken from the soil most efficiently for plant growth.

This increased mobilization and exchange of the nutrients into the plant, by means of the active hydrogen originating around the root from its excreted respiratory waste of carbon dioxide, is nature's way of getting the most plant growth from the little fertility left in many soils. Soil acidity is therefore beneficial. Any excess of limestone that would keep that acidity from so serving is one of possible sources of trouble in the health of our animals. Excess calcium carbonate in soil may mean deficient mobilization of the many other essentials from soil into plant. It reduces the amount of potassium taken and unsuspected symptoms of this deficiency occur with reduction in crop. This deficiency is serious because it may carry through several seasons. It also reduces the amounts of the trace elements, copper, zinc, manganese and boron active in moving into the plant roots.

Such deficiencies mean deficient proteins in total and in quality. This may mean deficiencies in the proteins of the genes in the chromosomes and thereby possible loss of some character commonly transmitted from parent to offspring in the flow of the life stream. Possibly by noting the ecology of some serious animal ailments in relation to the fertility of the soil as they are natural or have been brought on by man, there may come possible suggestions for prevention. We may find evidence that we are bringing them on ourselves. Calcium is the premier element for plant and animal nutrition only when it, along with all the other essential nutrient elements and even soil acidity as hydrogen, a non-nutrient, are properly balanced to meet the nutritional needs of the particular plant species we grow.

ADSORBED CALCIUM PLUS SOME ACID IS THE INDEX OF THE SOIL'S DEVELOPMENT AND CLASSIFICATION

The calcium is the premier element in the soil in serving as an index item of the degree of soil development. West of the approximate 98th meridian of

Degree of Saturation By:		Calcium in Plants:		
Calcium %	Hydrogen % (acidity)	Percent	Total Mgms.	Efficiency of Movement Into Plants
25	75	0.27	40.27	40.2
50	50	0.55	85.54	40.7
75	25	0.71	122.40	40.8
	Barium % (no acidity)			
25	75	0.29	1.20	31.2
50	50	0.31	45.54	22.7
75	25	0.66	104.84	34.9

Available calcium in the soil is moved into the crop more efficiently when accompanied by hydrogen or acidity than by barium and no acidity. (Data by Hutchings. Missouri Agricultural Experimental Station, Reserve Bulletin 243, 1936.)

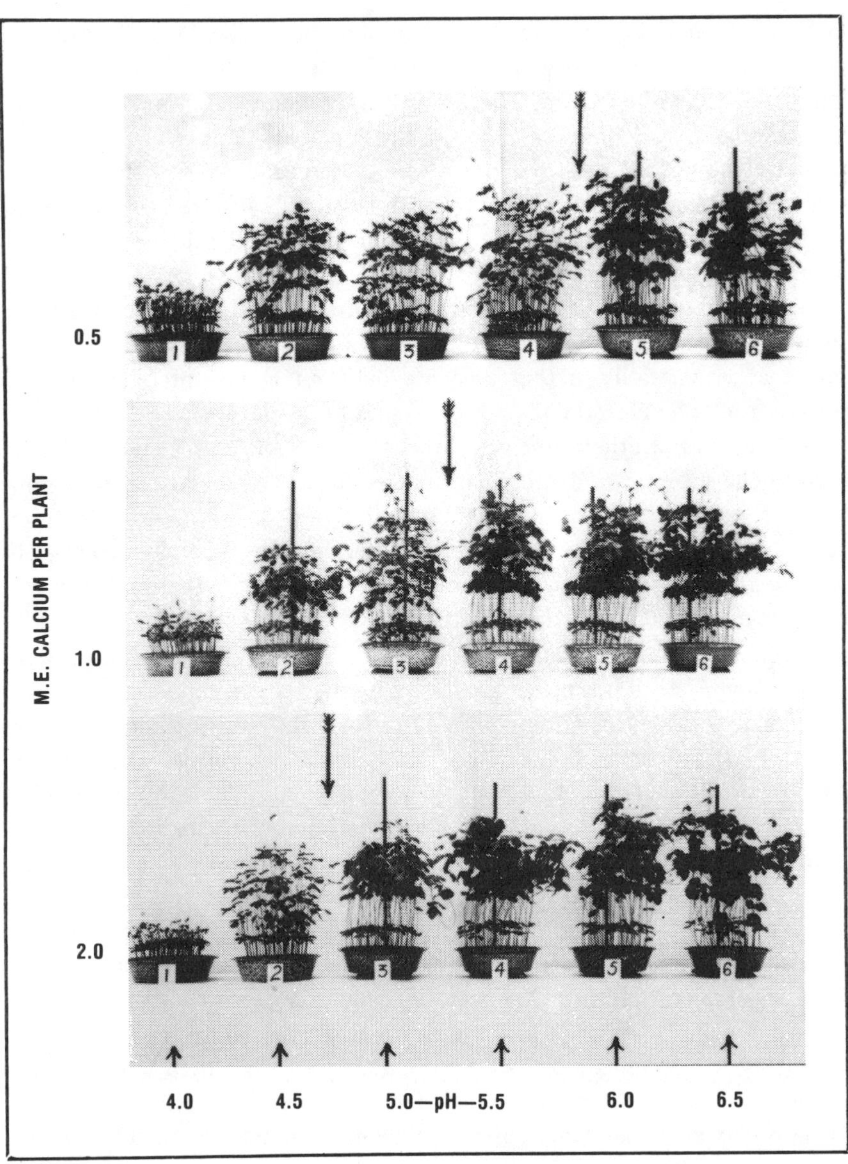

Putting more calcium-clay of any pH (going from the pans in the upper row to the lower ones) gave improved growth of experimental soybean plants. It demonstrated decreasing "sensitivity of plants to degree of soil acidity" or to pH as shown by the arrows. All of these sets of plants, except the lower right hand one, lost fertility back to the soil to indicate the "hay crop" we can grow which could not return a seed crop equivalent to the amount planted, nor could contribute as much of some nutrients for the cow's nutrition as if she had eaten the seed.

longitude there is a liberal supply of calcium in the soil profile. This is the index of only a moderate degree of soil development. Soils near that area are well saturated with calcium. They grow legumes naturally. They are granular, deep, well-stocked with organic matter and are of dark color—all as properties combined to make them grow proteins readily. We say they are not acid, which they cannot be when they are so highly saturated with calcium, magnesium, potassium and other cations to be so fertile for production of crops of high protein concentration.

East of the midcontinent we say the soils are more highly developed. The shortage of the calcium adsorbed on the clay is the indicator of this degree of soil weathering. Soils in that area are deficient in this premier element because under higher rainfall and more carbonic acid from decay of more organic matter grown there, the acid hydrogen has come on the clay colloid to replace the calcium and other nutrient cations. Those soils do not grow legumes naturally. They are more compact, of less granular structure because of more clay, have less organic matter, and are of a lighter color—all to make them less productive of protein-rich plants though they may be highly productive of the carbonaceous ones.

Calcium is the index of soil classification since many other elements run in its company. Thus if it is present in the soil profile we can expect many others to be there too. If it is missing or low in supply they may be anticipated in short supply too. Liming alone on a highly developed soil is, therefore, no remedy for the shortage of elements in company naturally with this premier of the fertility crowd. This natural service by calcium of indicating the higher or the lower productivity outlines the degrees of soil development according to the amounts of rainfall. It gives us our east and our west, and our midcontinent with all the adaptation of the crops according as their nutritional requirements are met by the fertility of the soil and not according to the simple comforts of the climate as to how wet, dry, cold or hot. For healthy animals it is the nutritional comforts which are required. These call for the soil fertility producing the proteins along with carbohydrates rather than only the carbonaceous crops making fat. The soil fertility pattern suggested by calcium as the index determines the pattern of healthy animals via good nutrition. Within this pattern the calcareous soils in the west and the non-calcareous ones to the east make calcium the major index of this simple classification. This is the case because calcium is the major inorganic element in animal nutrition and can come only from the soil via the crops.

**THE PROBLEM OF PROPER PROTEINS REVOLVES ABOUT CALCIUM
PROPERLY BALANCED IN THE SOILS TO GROW THEM**

The differing degrees of soil development register themselves in differing animal nutrition and health. This is illustrated by the report that cattle fed hays grown in various locations from east to west across Kansas ate less and less of mineral supplements according to that traverse. With decreasing rainfall and less development of the soil by that climatic factor in going that direction, the soils have more active calcium and grow higher concentrations of protein in the wheat grain so that, seemingly, the hays satisfy the cattle better; or at least reduce their desperation to get calcium even by eating limestone directly from the mineral box.

Also another case of moving dairy cows of recorded high production from the temperate midcontinent to the warmer and drier Bahama Islands, with soils composed mainly of calcareous coral remains with little else of fertility but calcium, showed a marked decline in the non-fat solids (carrying the proteins) of the milk while its volume, the condition of the hair and other indicators of health seemed to be holding up well. Possible changes in the chemical composition of the blood were not recorded. These illustrations and others tell us that the calcium, by either its deficiencies or its excesses as an indicator of the higher or lower degrees of soil development, tells much in terms of animal health if we know whereof it speaks.

That calcium is the premier of nutrient elements is suggested by its role of mobilizing those other nutrients in its company in a fertile soil from there into the plant roots. This seems to be due to its function in the inter-cell membrane and through possibly the higher protein concentration of the cell's contents. Legume plants on soils low in calcium may be growing to multiply their vegetative bulk. They may be making hay crops but not seed crops. But they may contain less of total nitrogen as protein, less of phosphorus and less of potassium than was present in the planted seed from which growth started. Plants seem to struggle desperately in making growth to survive, even to the point of losses to the soil of some of the seed contents whereby the growing of the crop may be a case of soil fertilization from that source. But as far as nutritional service to the cow by that crop is concerned, she would get more protein and more phosphorus should she eat the planted seed rather than the entire crop including the plant roots.

Our failure to comprehend the facts about the soil's calcium requirements to grow crops with nutritional rather than only filler values in them means that we are deceiving the cow and ourselves. That is the case when we believe the forage has feed value because it grew and made bulk of many fold of the planted seed. When "to be well-fed is to be healthy," that does not mean that "to be well filled" is the same. It may be the opposite. The studies of the soil's variable calcium supply, in relation to variable nutritional values of the crops grown thereby, emphasize the deception to ourselves in believing crops to be feed because they grow. Unfortunately the deception to the cow is still much greater via her growth, her protection against disease, and her reproduction.

CALCIUM DESERVES FIRST CONSIDERATION IN SOIL MANAGEMENT FOR BETTER ANIMAL NUTRITION AND HEALTH

In our meeting the soil's need for lime, we have been slow to see the nutritional significance of this soil treatment for the health of our animals. Perhaps the simplicity of the operation of grinding a soft, naturally common rock and distributing it in crudely defined dosages over the field at low costs, with generally beneficial effects on the crop, has left us indifferent to its effects on the animals fed thereby. We seem reluctant to classify limestone as the premier fertilizer, which it is, since its effects on nutritional values are not so startling as any on yields per acre. We have emphasized its effects on the soil when fertilizers are generally defined for their improvements in the crop yields. Unlike starter fertilizers, limestone is not extolled so much for its first year returns on the investment. It is not distributed in well-labeled bags nor under a state inspection service for close guard of its guaranteed chemical composition. Apparently it is too common and too simply handled to be fully appreciated for its fertilizing services to both the soil and crop.

Such limited recognition of the importance of calcium in our soils for healthy livestock would not prevail were we comprehending its fuller functions. Unfortunately the early encouragement for its use rested on the belief that the beneficial function of calcium carbonate applied on the soil rested solely in its reaction with the soil acids to convert their hydrogen into carbonic acid. This resulting acid decomposes quickly to leave water in the soil while carbon dioxide escapes from there. Hence a soil formerly with acid now is one with water in its place. No one was curious enough to test other carbonates, like that of sodium for example, to see if it had similar effects on the crop in consequence of its reducing the soil acidity. Applying the carbonate of sodium hinders rather than helps crops on acid soils. Failure to test the acidity as cause of crop failure by such tests using the plants left limestone use to continue these many years under the erroneously emphasized function of fighting soil acidity. Soil acidity was believed the soil condition detrimental to the crop when it was the absence there of sufficient calcium as a nutrient for plants and animals.

We need only recall what Benjamin Franklin demonstrated on our highly developed eastern soils by using land plaster, or gypsum, to spell out his report in the shape of greener letters marked out by the better red clover in the field. He showed us that while this sulfate of calcium was making the soil more acid by its sulfuric acid residue, it was also feeding the clover more calcium as nutrition to improve the failing crop. In similar manner, it was only necessary in some Missouri experiments to treat the soil with calcium chloride to grow better legumes by the calcium so added, even if there was the resulting muriatic acid as extra acid in the soil.

Failure to demonstrate the fact that the soil acidity was not disturbing the crop plants when they were well nourished by plenty of calcium (and all else) left us believing that soil acidity removal was the reason for liming with a carbonate of calcium, even when other chemical combinations of calcium may serve as well. Laboratory gadgets measuring the changes in soil acidity rather than measures of the changes in nutritional values of the crops encouraged the use of limestone. These are some of the reasons why calcium has not had our recognition as the major fertilizer element for growing the more nutritious forages.

LIMESTONE IS THE MAJOR FERTILIZER FOR BUILDING UP THE SOIL'S SUSTAINING FERTILITY

Limestone may not be considered a starter fertilizer. But it is the major

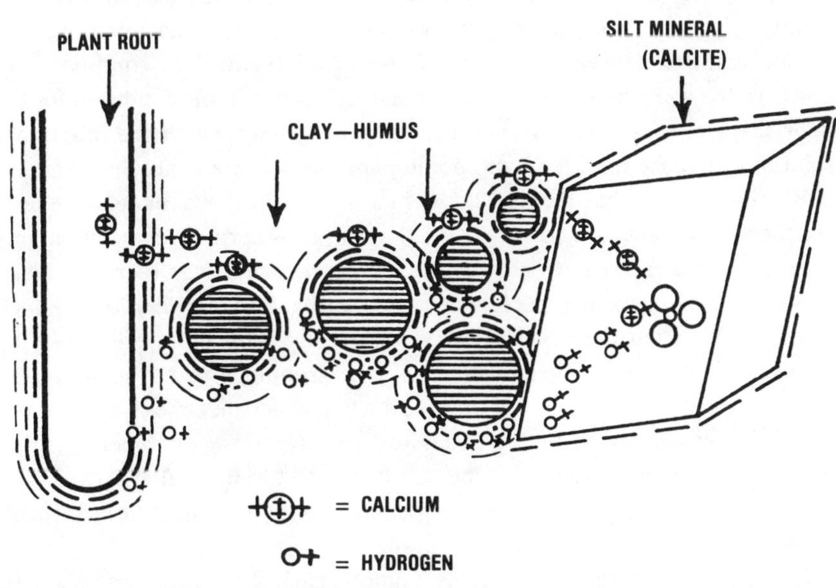

In the mechanism of plant feeding, the hydrogen around the root (originating in its respiration) is exchanged to the clay colloid or humus colloid for cations like calcium. As these colloids become exhausted of nutrients, or saturated with hydrogen, the hydrogen or acid moves to the silt mineral to decompose it and set its nutrient contents, like calcium, into active movement along the same line but in reverse direction from the mineral to the root for plant nutrition.

Calcium, the Premier of the Soil's Nutrient Elements   49

means for restoring the sustaining fertility of our soils. We have not classified it with the starter fertilizers, that is, the commercial, concentrated salt fertilizers because of the delayed crop response to its application. Many commercial fertilizers carry calcium. This fact is not reported on the bag even though calcium may be contributing beneficial effects on the crop from using that combination. We have failed to realize that nature's method of feeding crops consists in using natural rocks as combinations of minerals on which the acidities of the plant roots and of the soil react to make the nutrient elements therein become active and available for plant use. This involved no great amount of highly soluble salts put into the soil at one time. It avoids the possible salt shock to the microbial life of the soil and the injury to the germination of the seeds, or to the emergence of the seedlings which hazards are becoming more common on sandier soils or those going lower in organic matter when persistently treated with the highly concentrated starter fertilizers.

Agricultural limestone is in part a quickly-active fertilizer in its finest fraction. It is also a sustaining calcium fertility in the remainder of its varied sizes of particles. The former reacts quickly to change the soil's pH, or degree of acidity. The latter, or coarser particles, act but slowly as focal points of extended calcium delivery as the plant roots find them. This condition emphasizes the soil's heterogeneity as the natural and suitable condition for the nourishing of plants rather than a homogeneity with which the laboratory solutions or those used for hydroponic plant growth are characterized.

When we recall that the buffalo was natural on the soils made up of lessweathered minerals, and that such soils grow protein-rich or highly nutritious grasses, we may well ponder the practice of using other pulverized rocks besides limestone for soil restoration in sustaining fertility for livestock production. When the cow shuns forages fertilized no more excessively by nitrogen than by the concentration of it in droppings of her urine in the pasture, the question may well be raised whether fertilizers of nitrogen and potassium concentrated highly enough to threaten stands of wheat seeding would make choice forages for the cow were she given an alternative. How nutritious such feeds are needs wider test and demonstration through bioassays by some of our domestic animals.

Using rabbits as the test animals, some experiments have demonstrated their most uncanny discrimination between grasses grown on soils given different rates of nitrogen fertilization. They discriminated similarly with corn grains. Using sheep, the hays grown on soils fertilized with phosphates alone failed to produce a sound, cardable wool fiber. But an excellent fleece and fiber, stable under scouring and carding, resulted if the same plant for feed as hay was grown on the same soil fertilized with calcium as well as phosphate.

Using the above hays as feed for male rabbits serving in artificial insemination, those males fed on the hay grown on soils given phosphate only soon became sexually impotent as shown by the reduced volume of semen, lowered concentration of sperm and lessened viability of it. Those fed the hay grown on the soil given both the phosphate and the limestone remained potent. By interchanging the hays for the rabbits, the situation as to sexual potency was reversed in the short period of three weeks. These facts emphasize very forcefully the significance of the soil calcium as it reflects its chain of effects, extending from the soil dynamics through the plant's physiology and through the animal physiology, into the procreation and survival of our species of livestock.

KNOWLEDGE OF SOIL FERTILITY FOR ANIMAL HEALTH
MEANS BETTER DIAGNOSIS FOR DISEASE PREVENTION

By following the significance of calcium in the soil for the creation of the proteins, the life-carrying substance in the microbes, the plants and the animals, it has become evident that the chemical composition of the cow's blood and other details of that blood picture reflect the fertility of the soil. With the cow's blood varying more widely in chemical details than human blood, calcium has long been a premier therapeutic agent administered through the blood, as milk fever treatment demonstrates. It is not too great a

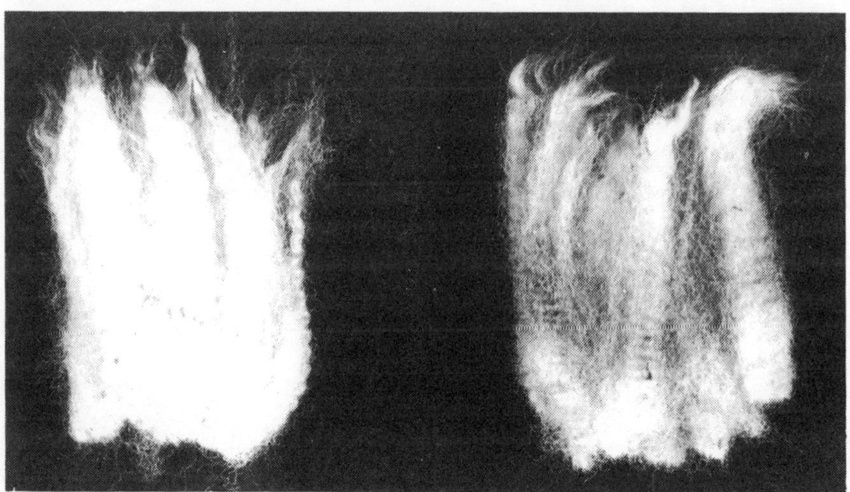

Limestone adding its calcium to the phosphate as soil treatments for growing the hay fed the sheep produced wool rich in yolk and of cardable fibers (left) in contrast to dry, noncardable wool from sheep fed on similar hay, but grown with phosphate only on the soil. This wool could not be carded, but broke up into dust (right).

stretch of the imagination to relate the chemical and biochemical details of the cow's blood to the forages she eats and thereby to the soils that support her.

Some experiences so far suggest that studies of blood composition relating it to soil test results may interpret animal health irregularities and may suggest their prevention by soil treatments. Knowledge of blood deficiencies connected with soil deficiencies would be helpful, even though we might not be able to interpret the significance of small variations in amounts present. Such interpretations of the relations between soil fertility and animal blood suggested and helped to demonstrate that dwarfism can be prevented by ministrations even after pregnancy and contradicted the contention that this kind of degeneration of our livestock is inherited. The soil deficiency served to bring calcium into focus in the larger picture of the degree of the soil's development and the fertility problem by which complete proteins are synthesized in our crops. We are but slowly realizing that the soil fertility is the basis of our animal health.

CHAPTER 6

Magnesium and Some Other Neglected Fertility Elements

FIGHTING SOIL ACIDITY with the carbonate of calcium has delayed, by many years, the appreciation of both magnesium and calcium as nutrition for plants. Magnesium is next to calcium in the order of the larger amounts of ash elements required in the soil's fertility store. While potassium makes up the largest amounts in the ash of plants, 2.14%, magnesium makes up but .31%, and calcium, as a mean, composes .88%. In animal bodies, calcium represents 1.50%, potassium .35% and magnesium only .05%. However, in order to grow the more nutritious plants, the calcium must represent 75 to 85%, magnesium 10%, and potassium but 3 to 5% of the saturation capacity of the soil's colloidal complex. This illustrates the higher degree of saturation of the soil's capacity by calcium and by magnesium than by any other element, even than by potassium of which more, in total, moves from the soil into the plant than of any other element. By fighting soil acidity with the carbonate of calcium, we have unwittingly built up the soil's saturation by calcium but have neglected the soil's need for magnesium. This neglect of magnesium under the widened ratio of calcium to magnesium in the soil is the more serious because while this widening of it does not lower the crop yields so noticeably, the concentration of magnesium in the forage grown depends, to a large extent, on this ratio with less magnesium in the plants as the ratio of them in the soil is wider.

Some few earlier students of soils suggested that dolomitic limestone, *i.e.*, the double carbonate of calcium and magnesium, should be used on acid soils since it is more effective in respect to acidity by 9% than is the pure calcium stone. But the greater hardness in grinding and the slower rate of reaction of dolomite in the soil kept it in bad repute economically as the limestone for soils. Calcium limestone remained the favorite. This discrimination aggravated the neglect of magnesium, increased the depletion of the soil's supply of it and magnified the imbalance or the very wide ration of

Chemical Composition of the (a) human body, (b) vegetation, and (c) soil, with the sources of the essential elements.

Source	Elements*	Human Body % Weight	Vegetation % Dry Weight		Soil† % Dry Weight
Air and Water	Oxygen	65.00*	42.9 (2)		47.3
	Carbon	18.00	44.3 (1)		.19
	Hydrogen	10.00	6.1 (3)	.22	
Air and Soil	Nitrogen	3.00	2.63 (4)		
Soil	Calcium	1.50	.88 (6)	0.3‡	3.47
	Phosphorus	1.00	.34 (8)	0.0075	.12
	Potassium	.35	2.14 (5)	0.03	2.46
	Sulfur	.25	.30 (10)		.12
	Sodium	.15	.70 (7)		2.46
	Chlorine	.15	.70 (7)		.06
	Magnesium	.05	.31 (9)		2.24
	Iron	.004	.0251 (11)		4.50
	Manganese	.0003	.01 (12)		.08
	Iodine	.00004	.00004		—
	Copper	Very small amount	.0011		—
	Zinc	Very small amount	.0041 (13)		—
	Fluorine	Very small amount	.0005		.10
	Aluminum	Very small amount			7.85
	Boron		.004 (14)		—
	Silicon				27.74

*Order of magnitude

†Collected from various sources.

‡ % readily exchangeable in soils.

calcium to magnesium. This was a still more serious neglect of this latter essential fertility element, which is so closely similar chemically to calcium that these two precipitated in conjunction as the double carbonate or the dolomite in the prehistoric sea.

MAGNESIUM IS IN THE "CATALYST" CLASS

The excessive applications of calcium in limestone as soil treatments, under the belief that if a little is good, more is better, and the encouragement by the low costs of this soil treatment, brought on the imbalances of calcium not only in relation to the magnesium but also to some of the trace elements. They and magnesium are much in the same functional category or in a similar classification in plants and animals. This is suggested by the small amounts of magnesium in plants and especially in animals to list it in, or at least close to, the trace elements class. In this functional grouping, the magnesium and the trace elements are not so much the materials of construction in the plant, but rather are the tools bringing that about. They are used repeatedly within the plant processes in causing elements and compounds to become larger and more active ones. They do so much by their own very small amounts. They are, therefore, mainly in the catalyst class.

It is in this catalyst class that magnesium serves when it is the core of the green chlorophyll of leaves catalyzing the process of photosynthesis. Therefore their magnitude as function transcends the amount of them in the ash relative to the mass of growth produced. While one hammer and one saw are required to build one chicken coop, the building of a hundred of them does not require equal multiplication of this number of tools. The magnesium and trace elements as tools in plant and animal nutrition required at the very outset are not multiplied or increased at the rate equivalent to that of their growth.

Dropped in August of a drought year with a bad winter, even a noble pedigree let this calf go down in April with a right femur telescoped into itself near the hip. The second femur broke while the heifer was being brought outside the barn for the photograph. Feed grown on acid soil with no treatment provoked the owner's report, "I don't seem to have any luck with my livestock."

Magnesium demonstrates its service as the catalyst in the green chlorophyll of leaves by causing carbon dioxide and water to be reduced, to give off gaseous oxygen, and to result in the simple sugar. Six units of each of the starting essentials for photosynthesis by plants give off six units of oxygen. This reaction forms one unit of a six-carbon sugar. This is a highly efficient process. About 30% of the sun's energy is caught and stored in this compound to be released for energy in the processes of life in both plants and animals. When this sugar is burned or oxidized in the plant and animal functions requiring this energy release by respiration, the reaction is exactly reversed. Thus the sun's energy, trapped by the sugar through the catalytic services of the magnesium containing chlorophyll in the plant, is released for service in the processes of life. But these resulting living products represent only about 2% of the sun's energy, or a relatively low efficiency.

Chlorophyll is unique in its chemical structure in that its central core consists of one molecule of the inorganic element magnesium. Around it are the compounds of nitrogen, suggesting protein as next. Then still around those are other organic compounds suggesting vitamins next, also now known to be mainly of catalytic services in life processes. Magnesium is only 24 parts in the total of 900 or more parts of either of the two kinds of chlorophyll.

A THEORY PROPOSED FOR ONE OF MAGNESIUM'S SERVICES

The element magnesium has recently become appreciated as part of many other catalysts serving in the plant and animal processes. The medical arts have long been using magnesium salts, especially the sulfate form. That we have understood how either the magnesium or the sulfur serves in the medical arts is doubtful. Theoretical consideration may well be given here, then, to the significance of both the cation magnesium and the anion sulfate when Epsom salts or magnesium sulfate, for example, are used as a purge of the intestinal tract, as internal fluids, and even as hot solution poultices. Their uses have long been an art. The science of their service seems to be still much of an unknown.

In the light of newer knowledge of colloidal chemistry, the purging effect may be viewed as one in which the magnesium enters the cells of the intestinal wall to exchange itself for the calcium in the so-called semipermeable membrane wall, and thereby changes the permeability which normally lets the water, etc., move from the intestine into the blood stream to one which allows water, calcium, and other materials to move excessively in the reverse direction from the blood stream into the intestine to flush it and give the resulting purge. This purging may be expected to continue until the blood stream, taking up the magnesium, exchanges calcium to the intestinal wall or cell membranes. This biochemical restoration of the normal intestinal wall stops the purge or this reverse movement of water from the blood stream into the intestine. Contemporaneous with the purge, there is the uptake of generous amounts of magnesium sulfate by the blood stream.

This fact raises the query whether the improved health effects result from the purge *per se* or from the advent into the blood stream and into the body metabolism of the extra magnesium and sulfate. Entrance of these through absorption by the skin with a hot poultice of Epsom salts may raise the same question, leaving the functions of both the magnesium and the sulfur of this compound still in the realm of the arts and not yet carefully catalogued as a science.

FUNCTIONS OF MAGNESIUM IN PLANTS ARE INTEGRATED
WITH FUNCTIONS OF OTHER ELEMENTS

It was such theoretical considerations as the foregoing which encouraged research studies of the role of magnesium in plant nutrition, with this element adsorbed on the colloidal clay of the soil and exchanged into the plant root for acid, from there in the plant's production of proteins in the case of legumes, like the soybean. Since the amount of magnesium by ash analysis is so low in animal body (.05%) and in the dry matter of the plant (.31%), we have possibly been slow to credit the service of magnesium as one in connection with the plant's production of proteins. The enzymes, or the catalytic tools in the plant's chemical reactions of which so many contain magnesium, are themselves proteins and are serving to bring about changes of both the carbohydrates and the proteins. When connected with magnesium many proteins become catalysts, but without the magnesium they are not. They are said to be denatured when the magnesium is removed. Magnesium is said to be a coenzyme, or a prosthetic element in this situation, determining the enzyme's possible service.

We have not seen large increases in vegetative production on many soils resulting from the application of magnesium alone. It is doing its work more with proteins which are so commonly deficient because of more common shortage of other elements required in much larger amounts in the plants and animals. However, now that we are using more nitrogen in the hope of building up the proteins and are beginning to study the production of proteins of nutritional quality along with the production of carbohydrate as bulk and energy, the shortage of the magnesium may well be one of the many hidden hungers not exhibited so readily by the decreased tonnage of bushels of yield. But the body's protection against degenerative occurrences may be failing and its process of reproduction may well be declining unless the nutrition in terms of magnesium is well balanced, be it nutrition of plant or animal.

In some research on the possible functions of magnesium in the growth

The addition of magnesium chloride in the early life of the soybean plant is not as demonstrative in more growth as is that of calcium chloride or acetate. Both are taken during all of the plant's growth period, but calcium shows marked effects early.

processes of soybeans by increasing the exchangeable magnesium offered by the soil, Dr. E. R. Graham demonstrated the influence on the movement of increasing shares of the soil's exchangeable calcium into the plant tissue. When smaller but increasing amounts of magnesium were used, the shares of the soil's constant exchangeable calcium supply mobilized into the crop were 31.2, 37.5 and 42.3%. When the amounts of magnesium were higher, the corresponding values were 18.7, 21.0 and 27.0% respectively for the portion of the available calcium mobilized by the magnesium.

It is significant to note that the soil's supply of calcium was removed by over 40%, or one-third higher, as the result of more magnesium. At the same time, the percentages taken of the extra magnesium offered were .0, 60.2, and 71.2 respectively for the lower rates of magnesium applications and .0, 66.1, and 68.5 respectively when the rates of exchangeable magnesium used were higher.

In the succession of three crops of soybeans on the same soil, the exhaustion of its magnesium was shown when the concentrations of it in the plants went down rapidly. It went down relatively more rapidly as the total amount offered by the soil was larger. With the three lower levels of increasing magnesium offered in each series and in the three successive crops, the percentages of magnesium in the dry forage were (a) .13, (b) .27 and (c) .47 for the first crop; (a) .08, (b) .10 and (c) .12 for the second crop; and then (a) .12, (b) .12 and (c) .08 for the third crop. With the higher levels of magnesium offered by the soil, the percentages of magnesium in the dry plant tissue were (a) .12, (b) .55 and (c) .86 for the first crop; (a) .09,

(b) .23, and (c) .39 for the second crop; and (a) .09, (b) .13 and (c) .27 for the third crop. All of these suggest that the plant composition soon reached a low level, around .10%, apparently required in these trials to give any plant growth. Also there was nitrogen fixation in the first crop, but there was little or no nitrogen fixation as the magnesium of the soil was more nearly exhausted and the concentration of magnesium in the crop moved toward this lower figure.

The similarity in the behaviors of the magnesium and calcium, in that they are both moved into this legume in relatively large percentages from the soil's supply as the soils naturally have them, is worthy of note. Yet because magnesium is in such a small percentage of the plant as a whole, we have not appreciated the protein-producing legumes as plants requiring higher levels of magnesium as well as of calcium in the soil.

CAN NEGLECT OF MAGNESIUM
BE THE CAUSE OF DISAPPEARANCE OF RED CLOVER

The higher concentrations of calcium and magnesium in legumes like red clover and alfalfa as compared with those for the non-legume timothy, and the differences in concentrations of calcium and magnesium between these two legumes offer suggestions and provoke some questions. Timothy hay—not considered good cattle feed—has but one-fourth as much calcium as red clover and one-fifth as much as alfalfa. It has less than one-third as much magnesium as red clover and less than one-half that of alfalfa. When red clover is more than three times as rich in magnesium as timothy and more than twice as rich in this essential element as alfalfa, should we not suspect that very probably it is through the neglect of magnesium in our soils

Chemical analysis of Timothy, Red Clover and Alfalfa (a)
(Figures as percent of dry matter)

Kind of Plants	Ash	Calcium	Magnesium	Potassium	Phosphorus	Sulphur	Sodium	Iron	Silicon	Chlorine
Timothy	6.82	0.39	0.13	1.76	0.35	0.064	0.092	0.039	1.02	0.35
Clover (b)	6.86	1.71	0.44	1.75	0.38	0.073	0.01	0.051	0.086	0.25
Alfalfa (c)	7.38	2.22	0.21	1.44	0.27	0.131	0.10	0.095	0.32	0.22

(a) Recalculated from USDA Yearbook, 1938, page 781.
(b) Clover in bloom.
(c) Alfalfa in early bloom.

that the red clover is about extinct, not only as the crop reports have it, but also as the extremely high price of red clover seed suggests? Can the neglect of magnesium be reason for the replacement of much red clover on limed soils by alfalfa, still possible in spite of our failure to be concerned seriously about magnesium as a fertilizer by way of dolomitic limestone instead of only the calcium stone?

It is significant that in Ohio, Pennsylvania and some other areas with soils from dolomitic limestone, there is still much red clover grown. Perhaps red clover in combination with timothy hay was such good feed because of its unusually high concentration not only of magnesium but also of many other of the fertility items. Have irregularities in animal health, like dwarfs, shown up in herds fed on red clover hay grown on soils from dolomitic stone or soils high in active magnesium? Perhaps some ecological surveys of these health irregularities of cattle in relation to soil fertility, especially magnesium, will offer suggestions as to causes and thereby help for prevention rather than cure by killing the unfortunates.

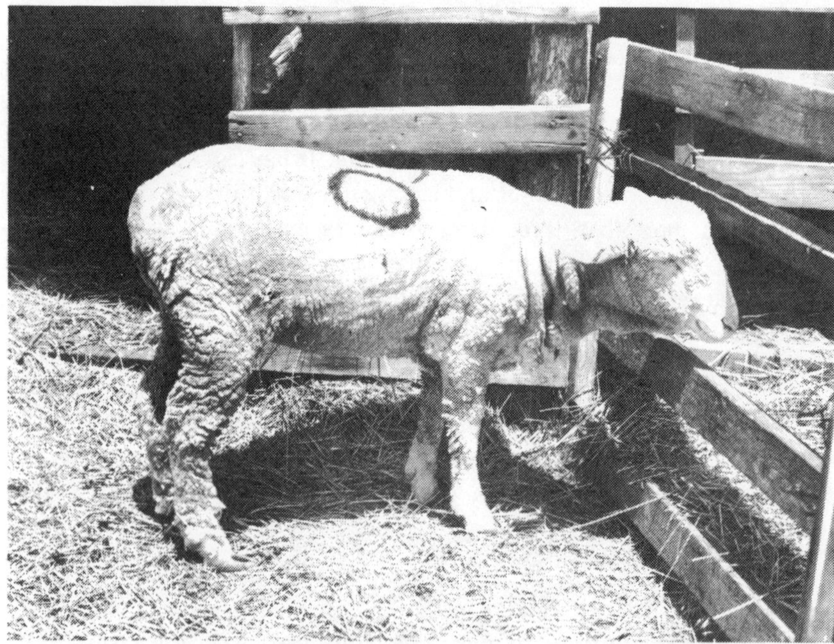

A period of experimental winter feeding on timothy hay, grown on the level Putnam silt loam with its deficinent clay subsoil but given no soil treatment, made some sheep get "sick," go lame in one hind leg, lame in both, "go down," have convulsions and then die. Crops may make bulk but not feed on so-called "acid" soils where "liming the soil" must be a fertilizing practice adding magnesium often as well as calcium.

We have not appreciated the interrelations of one element to each of the others in their movement as a group, or a suite, into the crop from the soil. We have emphasized one as the limiting element so much that we have not seen other elements in such low supply that they become limiting elements soon after we add generously the first one we recognized. Adding the limiting element generously without considering other elements in supply, as we do calcium in excess for disturbance to magnesium movement into the crop, may be an erroneous and dangerous practice.

Adding potassium liberally as fertilizer is another case of so-called antagonism to the movement of magnesium into the crop. This is also true for the reverse, namely when excess of magnesium put on the soil brings on re-

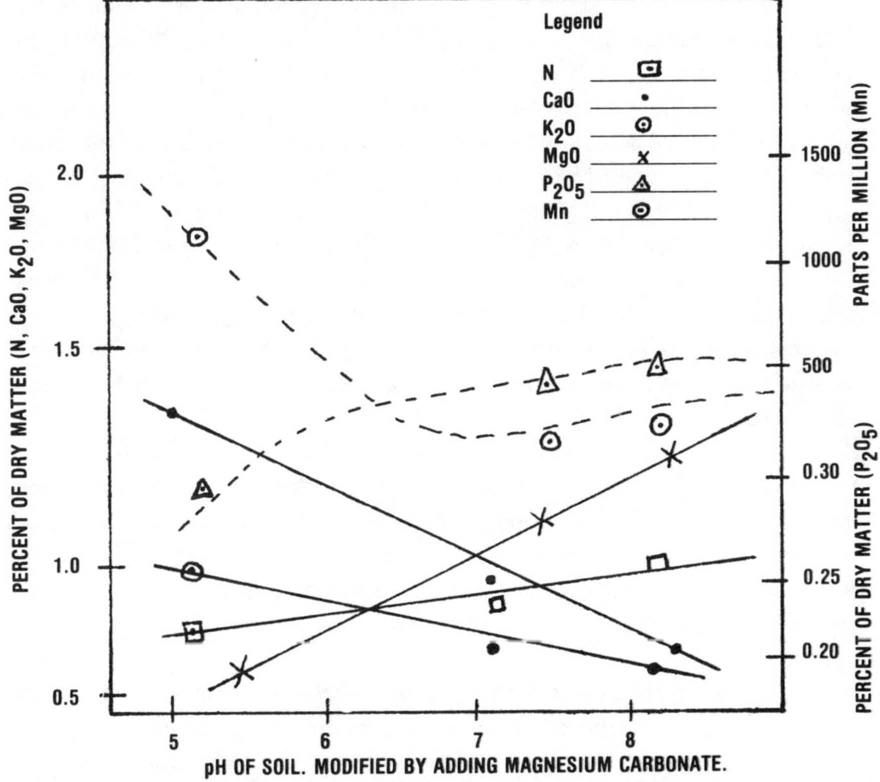

Using increasing amounts of magnesium carbonate in place of calcium carbone to reduce the soil acidity (to increase the pH) moved (a) increasing amounts of nitrogen, N, magnesium, MgO, and phosphorus, P₂O₅, but (b) decreasing amounts of calcium, CaO, potassium, K₂O, and manganese, Mn, into the rhododendron plants from the treated soil. (Chart courtesy of Dr. Tod, Edinburgh and East Scotland College of Agriculture, Miscellaneous Publication No. 164.)

Magnesium and Some Other Neglected Fertility Elements 61

duced concentration of potassium in the crop. With the increasing use of potassium fertilizers, some irregularities in animal health should not be a surprise because of feeds inducing magnesium deficiencies. Excessive calcium from overliming the soil brings on deficiencies in the crop for its own better growth of also the trace elements, boron, zinc, and manganese. These are additional cases emphasizing the need to view the nutrient elements as an integrated group behavior rather than as the addition of individual effects in a succession of them.

SOME SYMPTOMS OF MAGNESIUM DEFICIENCY IN ANIMALS VIA PLANTS AND SOIL

Magnesium deficiency in cows and horses as impending cases of tetany or grass staggers usually goes unnoticed until the pronounced symptoms in the late or near fatal stage are evident. If recognized early, the feeding of magnesium salts is prevention. This occurs usually a week or two after the animals, especially high-producing cows, begin grazing on new grass in the spring. This is the season (May and June) when the magnesium concentration of the pasturage was lowest according to studies both in Great Britain and the United States. This affliction is now considered a case of hypomagnesia, as shown by the level of magnesium in the blood serum. Feeding rations deficient in magnesium or keeping calves on only milk too long lowers the magnesium in the blood. More common examination of the level of magnesium in the blood serum is a good diagnostic help now that so-called normal values for magnesium in rats, cattle, humans, and others have been established.

Magnesium as it functions in enzymes as tools may well remind us that since those tools are proteins or nitrogenous organic substances compounded with the inorganic element magnesium, we do not see the deficiency of magnesium reflected in carbohydrate production or bulk increase of plants so readily. Rather, the magnesium reflects its importance in connection with the proteins by which all life forms (a) grow, (b) protect themselves against disease or degeneration, and (c) reproduce.

The essentiality of magnesium in the soils and feeds as protection against microbial troubles in the intestinal tract as white scours and in the lungs as pneumonia, brought itself into strong light by some observations of dairy cattle on the Brookside Farms near Knoxville, Ohio several years ago. A new calf and maternity barn, built especially to improve sanitation in the struggle against the high mortality rate (23 to 41%) among newborn calves, gave occasion to have the sick calves eat the plastered walls and suggest the reason for their poor survival rate. The high mortality rate with young chicks had eliminated the poultry project of the farm. The swine breeding program had been abandoned because of the high mortality rate among the pigs. So-called animal diseases, because of poor animal nutrition via deficient soil fertility growing the farm feeds, was undermining the separate animal enterprises and thereby the business as a whole.

Even after concentrating on dairying, the mammary disturbances—milk fever and mastitis—in the milk cows, increasing shy breeders, and the loss of 49 calves one season out of a single crop of 120, as about the usual annual number, was reason for stalls with their walls plastered with two coats. Only a few of the stalls had the second coat applied and the remaining stalls had only the rough first coat when the fall weather put the calves into the barn. The mortality trouble continued. "Calves were born weak, with slow reflexes and no appetite; dietary scours developed in 100% of the cases; 50% of the cases were accompanied by a low type of pneumonia with much coughing; calves that died invariably went down with convulsions and no calf ever to go into the convulsion stage lived beyond six hours. Strong disagreeable odors were present in the stalls. Calves that lived through the first 90 days showed remarkable recuperative ability and matured to good size without any signs of these calfhood disorders."

The observation was then made that the calves were severely mutilating the walls which had been finished with the second coat of plaster. This provoked the query, "Why are the calves trying to eat the finish coat of plaster when the walls with only the first plaster coat are not mutilated or eaten?" Chemical analysis of this plaster used as finish showed that it was made from dolomitic limestone. The limestone's analysis gave near 54% calcium carbonate and over 45% magnesium carbonate or dolomitic limestone. With this came the suggestions that the calves know their own medicine and are capable chemists in discriminating between the finish coat of plaster made from dolomitic lime and the first plaster coat from a calcium lime.

This shifted the feeding practices to include dolomitic limestone in place of calcium limestone as mineral supplement to the grain ration supplement containing 16% protein. The soils were tested to find magnesium the lowest, or the limiting element and were then limed with dolomitic stone. In two weeks, after the changed feeding of the cows giving milk for the calves, the disagreeable odor in the barn was becoming less and there were changes in the droppings of the calves. There was increase in their thirst, they became alert, and the scours condition cleared from the older calves. After this improvement in animal health had occurred, the remaining stalls were given the finish coat of plaster with no single tooth mark or mutilation of the walls. Other evidence of better animal health was (a) the disappearance of mastitis which had been as high as 50%, (b) improved conception and (c) strong normal calves that did not go through the scours stage. Decided improvements in the crops resulted from magnesium additions to the soil. The corn plants were not so readily fired, the grain yields were increased and the alfalfa went through the winter better—all as evidence that the plants just as well as the cows, as a consequence, were suffering because of a deficiency of magnesium in the soil.

SULFUR ALSO A NEGLECTED ELEMENT

Sulfur, another neglected element, is effective also in very small amounts. It, too, is important via the proteins when methionine, the sulfur containing amino acid among the essential ones composing complete proteins, is so commonly deficient that its addition to crude proteins in very small amounts in the chicken ration lets the carbohydrate-protein ration of the ration be widened and yet makes a ton of feed grow more poultry. Also, treatments of the soil with sulfur as a fertilizer have increased the methionine concentration of soybean plants. Since the element sulfur in its reduced form in the nucleoproteins seems to be connected with efficient cell division—according to late thinking—the actual growth of plants and animals may depend on sulfur much more than is commonly recognized.

When elemental sulfur mixed with fats, like lanolin or petroleum, cures skin troubles and more deeply seated irregularities by external application of it, we may well raise the question how this supposedly insoluble element applied on the skin becomes available for absorption there to effect cures. Elemental sulfur dust inhaled by men working where it is pulverized does not bring on the equivalent of silicosis of the lungs. This fact suggests the ready absorption of this supposedly insoluble element by also the lung tissue. However, if the sulfate of calcium, much more soluble than the elemental sulfur, is inhaled where it is being pulverized, this brings on the equivalent of silicosis. Here the sulfate is a bit troublesome. Yet the sulfates of many elements used in therapy for animals and man are more effective forms than are the chlorides and the nitrates of them. These latter two as monovalents are not in the same category as compounds of the divalent sulfate.

It is well to remind ourselves further that it is not the oxidized sulfur, like sulfate in which this essential element occurs in the living body compounds, but rather in the reduced or S-H forms. Nitrogen represents a similar situation. This is in the tissue with the nitrogen connected to hydrogen, $N-H_2$, and not in the nitrite or nitrate (NO_2 or NO_3). In these latter two forms, nitrogen has been responsible for poisonings like the so-called cornstalk disease. This trouble may follow heat or drought periods which suspend the plants' enzyme action, and then after rainfalls bring the movement of excessive nitrate into the plant which is unable to reduce it because of the inactive enzymes denatured under the excessive heat. Excessive nitrates have been responsible for the deaths of some babies under unusual soil conditions and climatic settings putting nitrates into the shallow well water from which the babies' artificial formulas were made. Sulfur in the sulfate and elemental forms has long been an actor in the medical arts, but its functions in the body are not yet understood fully enough to move our knowledge of its behavior and use into the classification as medical science. But as an element of diverse properties and functions playing important roles in proteins for growth as significant as the division of the cell, this element goes along with magnesium as another much neglected element.

Sulfur in its elemental form has been used as a soil treatment with advantage on sulfur deficient soils. Insoluble as it is, the reaction of oxidization to give sulfuric acid soon takes place. Oxides of sulfur are brought to the soil by rainfall in amounts varying in relation to the coal consumption, but now that coal is no longer such a universal source of heat and power, the return of sulfur from that source is not overcoming its increased depletion from the soil. That, however, has been helped, unwittingly, when we use phosphatic fertilizers of, say 20% grade made by mixing a ton of commercial sulfuric acid with each ton of rock phosphate to supply up to 12% sulfur and up to 11% phosphorus in the mixture of more of the former than of the latter. Thus, the sulfur supply of the soil has been periodically renewed. Now that sulfur free fertilizer applications are heavier, crop removals are larger and soil organic matter supplies are rapidly declining. Sulfur as such an active and essential element in the proteins, and such a vital element in protecting health and aiding reproduction of warm-blooded species, dare not remain in the category of part of an art or as one of the neglected elements. By using bioassays or animal tests, the science of the soil undergirding plant nutrition is finding the biochemical significance of sulfur growing larger in animal nutrition.

MAGNESIUM IN BALANCE WITH OTHER FERTILITY ELEMENTS OFFERS HELP

Growing nutritious forages and feeds is more than a problem of carbohydrates balanced with crude protein. It is one of balancing the energy supplying feeds with proteins complete with respect to all the essential amino acids, and with respect, also, to the essential, major and trace inorganic elements more appropriately called bioelements or life-elements from the soil. Balancing the bioelements according to percentages in the ash, or in the dry matter, is not sufficiently accurate when the organic compounds of which they are constituents are nutritional essentials by their own organic right, more than because of their delivery of bioelements like magnesium, calcium, copper, zinc and others. Yet this does not reduce the significance of the bioelements of which many are not fed in sufficient amounts as ash essentials, even with that crude method of measurement.

A recent report by the United States Department of Agriculture cited two food essentials as the major deficiencies for humans in large areas. The foremost organic one they cite is vitamin C. The major inorganic one or bioelement they list is calcium. Perhaps when magnesium is studied as intensely and widely as calcium has been over these many years, its deficiency will even be more widely spread than that of its chemical mate calcium.

Magnesium enters into plants with varied amounts according to the degree of development of the soils under the climatic forces. On those less weathered or moderately developed by the moderate and low annual rainfalls, magnesium in the plants is of higher concentration. As soils are still less developed, calcium may so dominate the soil that magnesium is deficient in plants as animal forage. On highly weathered soils also, magnesium is often deficient. Plant composition followed a logical pattern and there are indications of trends of the behaviors of the elements by groups not only according to the soil but also to the period in the plant's growth. The ratios of the bioelement active in the soil and around the plant root are the major determiners of the kinds of plants most aptly fitted there.

Considering the period of the plant's growth, potassium for example enters early and shows a decrease in rate of entry near plant maturity. This suggests its dominant activity and use when increase in plant bulk by carbohydrate production is underway. High potassium in grass is common in grass staggers. As a general rule, magnesium and calcium enter at a fairly regular rate throughout the plant's growth. Phosphorus and copper usually show marked increase in rates of uptake from the soil by plants at increasing maturity, or when the biosynthetic products within the plant are being mobilized and concentrated in the plant's reproductive part like the seeds and storage roots. The regular flow of magnesium suggests the plant's regular demand for it during both the early photosynthetic processes of building the carbohydrate bulk or tonnage yield and the subsequent biosynthetic processes and products to guarantee the survival of the plant species by means of special proteins and all else associated with reproduction.

Faulty balance of the bioelements in the soil, including prominence of magnesium deficiency there, has suggested its possible causal connection with the problem of the dwarfs or the midgets in cattle. If we were to tabulate the scatterings of these cattle unfortunates (not widely announced) over the soil pattern of the United States and to couple the history of soil treatments with each of them, we would possibly be given suggestions that the problem is one originating in nutritional deficiencies, and these even traceable to the soil as fertility imbalances registered through the crop. When alfalfa, with its lower concentration of magnesium tolerates and is commonly given heavy calcium liming, there may be reasons in its disturbed chemical composition why this crop grows where red clover with higher magnesium concentrations as requirements will not. Certain orders of crop replacement resulting from declining soil fertility may be suggestive. If, with these locations of the reproduction failures to guarantee the animal's growth, we should study the crop composition correlated with detailed soil tests and should then correlate those data with the chemical composition and picture of the animal's blood, it is possible that an approach to prevention of such troubles like dwarfism and others about as perplexing would suggest itself. Degenerative diseases are telling us that nature's flow of the life streams of our domestic animals must soon have our help if increasing hog diseases, more disturbing cattle ailments and other irregularities are not to make the growing of livestock economically impossible.

For too long a time already have we allowed crops of higher nutritional values to be dropped out of our farming and feeding operations because we have accepted substitute crops rather than learn how to feed them via the soil and its fertility to keep growing the better ones. By those substitutes we have made the many feed supplements, including protein, mineral, antibiotic and others, a requisite. We are coming face to face now, apparently, with the threat of livestock species about to drop out of economic farming operations too when their offspring in high percentages is born without the capacity to grow. Unlike the case of crops dropping out of our farming scene, however, we have no long list of substitute animals for our present kinds of domesticated ones, especially none able to tolerate starvation for protein. We can grow them in health good enough for their survival only when the soils are fertile enough to feed them for it. That calls for a balanced fertility, both inorganic and organic, in the soil for nutrition of plants as well as balanced nutritional components of organic nature in the feed for the nutrition of the animals. The disaster, as serious as birth without the potential for growth which is now exhibiting itself in larger numbers, suggests that there are still many neglected fertility elements among which magnesium may be a major one in this particular trouble.

PLANT NUTRITION

Professor Miller thought I should grow bacteria that would make the cow have a calf whether she wanted to or not. And I had to politely show the points I wanted to make. Here [at which point Albrecht produced a report entitled, *Some Soil Factors in Nitrogen Fixation by Legumes*] is increasing calcium saturation of an electrodialyzed clay. I separated the finest part of the clay out in the centrifuge running 32,000 r.p.m. after the clay had been suspended and settled for three weeks. At the bottom, that clay plugged up finally because the clay was too heavy. But we had thinner and thinner, smaller and smaller clay until about halfway up in that centrifuge—you know what the milk separator is?—there we had it as clear as vaseline. Now we took the upper half of that clay—a clay so fine that it was like transparent vaseline. We made pounds and pounds of that because we had put it into the electrical field and made it acidic and took all the *cations* off so it was an acid clay. That was the thing with which we studied plant nutrition. We studied plant nutrition with that clay by putting different elements on in different orders. We grew plants. We studied plants with this fraction of the clay in the soil that holds the positively charged nutrients. And we could mix them and balance them.

We began with calcium because we found that we had to come up here to 65% saturation. In other words, you've got to load that clay in that soil with 65% of that clay's capacity to hold calcium against rainwater before you can grow a plant with enough calcium to be healthy.—*Dr. William A. Albrecht, in an interview with Acres U.S.A.*

CHAPTER 7

Phosphorus.
A Problem of Keeping
Enough of It Active

NUTRITIONAL DEFICIENCIES in our livestock, which occur so widely and are very important economically, do not necessarily show themselves promptly by clinical symptoms. They are often manifested through declining output of the animal's products or services before serious clinical evidence occurs. They are therefore hidden hungers. Deficiencies of phosphorus may well be the cause of cases classified under this category.

Shortages in the carbohydrate part of the feeds for supplying energy to the animal body are not commonly considered as troublemakers when many different kinds of them, like sugars, starches, hemicellulose, celluloses and others seem to serve interchangeably. But shortages of protein in the feed or forage have been a problem for a long time when the so-called commercial feeds had their origin in the need for extra proteins to supplement those of short supply in the feeds commonly grown on the farm. The term "protein supplements" has become common language. They make regular feed business for the animal in confinement and subjected more completely to the way we want to feed it for particular economic results rather than for the way the animal at liberty would feed itself for the physiological results its instincts for survival dictate. Under such forced feeding, duplicating as it were almost experimental laboratory conditions with more unnatural rations, should we not expect hidden hungers? The more recent but increasing diseases manifesting irregularities in the mucous membranes and in the skin—when both of these body parts are the same in terms of their embryological origin—tell us that these external organs of protection are breaking down or degenerating more commonly. We are also told that in more numerous cases of the liver, this internal, protective and chemical censor is also degenerating long before any symptoms suggest it.

We can no longer separate animal nutrition into a major matter of carbohydrates and proteins, with the latter measured by ashing the feed to determine its nitrogen and then multiplying that amount by an arithmetic factor to call the result *protein*. Neither can we emphasize the supposedly minor matter of vitamins nor of the inorganics in the so-called "minerals." There are neither *majors* nor *minors* in the requirements for nutrition. The deficiencies of some are just more quickly and clearly manifested as clinical symptoms than of the others. The numerous essentials, both as elements and as compounds of which many may be yet unknown, cannot be set apart individually for differing significance. They are all required for the life of the animal. The functions of giving energy, growing body tissues, protecting them from invasion and destruction through some foreign protein feeding on them and of reproducing a new body are not naturally separated and compartmentalized. They are integrated. They are all part and parcel of the same cell. Each element may be involved in many functions there.

PHOSPHORUS IS CONNECTED WITH THE FUNCTIONS
OF BOTH THE CARBOHYDRATES AND THE PROTEINS IN·THE CELLS

This integration is well illustrated by the soil-borne element phosphorus. We have long emphasized its shortages in supply and in activities in the soil. Its shortages in the soil occur because of the relatively large amounts going from there into the plant, especially towards its maturity, for the improved quantity and quality of the yield of seeds. Large quantities also go into storage roots, like Swede turnips, etc. and into other segments for reproduction in both plants and animals. But now that we are reminded that it is the proteins which keep the pilot flames of the body lighted so that the carbohydrates can be burned and give us energy, we have found phosphorus important in all that the carbohydrates can do as well as in all that the proteins can do. Then, when but little of the phosphorus in either the plant or the animal is required in a particular function, any minute irregularity in the active supply of phosphorus within a very short period may be a very serious hidden hunger, if not even a disaster.

Phosphorus is not so readily appreciated when it may be hidden away in some organic compound of which it is but a very small but most important part as the very core around which much that is organic is connected. The functions of this essential element of soil fertility are still not well elucidated when, for example, (a) it hides away in the bones, (b) it is only partially active for the supply in the circulating blood stream and (c) it combines with more than one element at a time to produce so many compounds so apt to be insoluble and inactive. Because of these properties which keep its presence and functions in the body hidden, we do not recognize its deficiencies quickly by any chemical tests in vogue. In our study of it, like for its services in the soil, the plant, and the animal, the problem is not only one of the presence of phosphorus, but also one of it in chemically active forms.

The activities of this essential do not involve the phosphorus in its elemental form. Rather, it functions in the combination—almost universally—of itself as one part combined with four of oxygen, or in the phosphate form. Even as this unit, it is commonly in further combinations with either hydrogen or calcium or the vitamin-like units of combined carbon, hydrogen and nitrogen. Thus, in general, this oxidized phosphate unit as an anion becomes highly important in its role in body processes using the carbohydrates as well as in affecting many other activities we associate with the proteins.

In building the skeletal part of the body we emphasize calcium, phosphorus and magnesium along with the carbonate combined with the calcium and the magnesium—as they are inorganics coming from the soil. But in building the plants via the common fertilizers, we have been emphasizing nitrogen, phosphorus, and potassium as soil treatments representing nutritional supplements for the growth of vegetative matter. Phosphorus is the one soilborne element common to both of these groups and serving uniquely in many functions in both the plants and the animals. It is essential in ever-increasing metabolic services as we find its catalytic and structural functions appearing in connection with the carbohydrates as well as with the changes involving proteins in the cells of both plants and animals.

While calcium is found in the bones and teeth to the extent of about 99% of its total in the body, save for some of it in the extra-cellular areas, phosphorus is also mainly in the bones in the human body, save that there is from 10 to 20% of the total in the soft tissues. These tissues seem to have priority over the bone for the active phosphate. Hence, bones may be giving up their phosphorus long before that sacrifice in this body part is recognized as a biochemical irregularity. For the metabolism of energy compounds, their combination with phosphate in a reaction called phosphorylation is an essential step for their absorption from the intestinal tract like the fats and the sugars. The phosphate radical is chemically bound to proteins, fatty acids, carbohydrates and enzymes, which fact illustrates the several food and feed components with which it may be chemically functional. It is tightly bound and the high energy involved permits gradual release of energy during oxidation, for example, like that from the liver sugar or glycogen in the muscle. The phosphate part of a compound as the anion or negatively charged ion is the chief negative one within the cell fluids. Outside of the cell the phosphate is essential in regulating the acid base balance of the body. Its excretion in the urine is a part of the mechanism of that maintenance.

Now that the use of chemical nitrogen as fertilizer for plants is more extensive and at more generous rates, the phosphorus shortages in the soil are magnifying themselves in seriously disturbed plant processes. These are suggesting imbalances in our soil treatments through such readily soluble and highly ionized, and thereby highly active, anions like the nitrate applied as such or resulting from the microbial oxidation of applied ammonia. Also there is the very active cation potassium coming into more extensive application on the soil. It is bringing along with itself the highly active and thereby also disturbing chlorine which is required in only trace amounts in plant nutrition. These in their highly dynamic ionic behaviors are all in decided contrast to those of the less dynamic phosphate.

As a trivalent anion the phosphate is more insoluble. It is not highly ionized. Consequently, it is not very active. Yet as an essential element, it must be made, such if it is to function, in balance with more active ones in both the plant and animal processes that build the body composed of protein and that oxidize the carbohydrates to give the energy to the cells of the bodies for the creative services they carry out.

The agricultural problem of using phosphorus, then, is not only one of having ample amounts of phosphorus present in the soil, the plant and the animal bodies as shown by ash analysis, but also one of increasing the activities of phosphorus in its inorganic form in combination with mainly calcium and hydrogen, and also in its organic combinations of many kinds. The levels of these activities must be high enough to make phosphorus literally be the fulcrum that determines how high it may be raised and how widely the other essentials may fluctuate and yet let all the growth processes be balanced well enough to undergird agricultural production.

THE PROBLEM OF PHOSPHORUS' ACTIVITIES IN THE SOIL

In the original rocks of the earth's crust, the phosphorus is a scarce element when viewed in relation to others for production of crops for animal feeding. Found in nearly all igneous rocks, it is there mainly in the form of the mineral apatite. While this mineral is widely distributed over the earth's crust, it is in relatively small proportions. There are some few places where it is concentrated into larger deposits or in veins. It is slowly dissolved in percolating carbonated waters. Hence, in the soil it is moved from its highly insoluble state to conditions of more activity by decaying organic matter giving off its resulting carbonic acid. Fortunately as a matter of conservation, it is thus held rather permanently in the soil. It is moved only slowly toward the sea. This is in decided contrast to other elements like calcium, sodium, potassium, chlorine, nitrogen, sulfur and others. In the presence of calcium carbonate, phosphorus is not readily dissolved, though again even apatite is more soluble in the presence of accumulated organic matter when swamp areas rich in humus dissolve this mineral and also the phosphorus in the simpler tri-calcium-phosphate. Moved to the sea in small amounts it is taken from there by the life in the sea. By the death of that, the phosphorus has been left in combination with calcium. Mineral phosphatic beds of calcium have been formed from literally the graveyards of fish. These have served as commercial sources of phosphorus for fertilizer use.

Fish collected from the sea by the cormorants and dropped in their roosting areas along the western shores of South America have made the guano deposits, once mined and formerly imported extensively into the United States as fertilizer nitrogen, a phosphatic fertilizer as well. When the guano was the first nitrogenous fertilizer used by the pioneer cotton farmer and was later replaced by Chile saltpeter, also from western South America, because of the emphasis on only the element nitrogen in these two fertilizers, it is not surprising that the southern pioneer preferred the guano. Doubtless he was cognizant of the extra values of the phosphorus from fish bones in the accumulated nitrogenous bird droppings, or guano. This says nothing of the organic compounds of fertilizer value directly, and also of value for their effects in making the phosphorus more active in this combination. Also the nitrogen in the organic form would be transformed by the soil microorganism more effectively when in company with the calcium phosphate of fish origin from the sea. Nitrogen transformations, in turn, make the insoluble forms of phosphate more active, too, as decaying green manures used in connection with rock phosphate fertilizer testify in agricultural practice.

In the very beginnings of commercial fertilizer production and use, it was the need for phosphorus that provoked and initiated what has developed into this present tremendous industry connecting itself with the provision of the fertility essentials needed for the soil. It was the low activity of the phosphorus in bones, initially used to supplement the plant's needs for phosphorus from the soil, that started the fertilizer business. Bones, as a form of tri-calcium-phosphate in an organic matrix, and in possibly other combinations of animal origin, were treated with sulfuric acid to result in calcium-acid-phosphate. This is a more soluble phosphorus. It is one ionically more active in the soil moisture. Later, the natural mineral forms of phosphorus, including apatite, a calcium-fluoride-phosphate, and the other natural deposits of phosphorus included in the general term rock phosphate were treated similarly. This was the major chemical process in the fertilizer industry, though it is less so recently with increased chemical fixation of nitrogen and the mining of potassium salts. Though there are other phosphate minerals besides those of the phosphates of calcium, including those of iron and aluminum, the insolubility of all of them, or their inactivity in the soil, the plant and the animal body, makes their use a problem of changing them enough chemically to mobilize into activity the inactive phosphorus they carry.

Phosphatic fertilizers divide themselves into two groups. These may be considered (a) the starter fertilizers or the more soluble and more active forms resulting from phosphatic minerals treated with acids to make an acid calcium salt of phosphorus, hence the name *acid phosphate* and (b) the sustaining fertility which is the highly pulverized natural phosphate minerals without chemical treatment applied in much larger quantities to the soil to build up the soil's supply of phosphorus. These natural minerals are intended to be slowly mobilized by the soil processes rather than to provide a highly active form used in smaller amounts with the seeding of the crop, as the acid phosphate is. Fertilizer production aimed its efforts to provide phosphorus of a high degree of activity. It has always been faced with the problem of the ready reversion of an active, or ionic, form of phosphorus back to an inactive form, or one not so soluble in but a short period after mixing within the soil. Emphasis has recently gone to combining the phosphorus with nitrogen to give ammonium phosphate—more soluble and more active than even much of the acid phosphate of calcium. All of this is more testimony that the problem of phosphorus is one of first, the generous presence of this element in the soil and second, one of keeping it ionically active there for entrance into the plant root.

CHEMISTRY OF SOIL PHOSPHORUS DIFFERS FROM ITS BIOCHEMISTRY

Paradoxical as it may seem, when we say that calcium in the soil may make them more soluble and active, calcium-acid-phosphate becomes less soluble and less active, so far as the chemist measures the activity of phosphorus; yet when phosphorus is applied as the calcium-acid-phosphate, it is mobilized into a legume forage crop, like lespedeza, in larger quantities per acre when the soil is first treated with calcium carbonate than when the phosphate is applied alone. This was demonstrated by some research studies at the Missouri Agricultural Experiment Station using (a) no treatment, (b) acid phosphate and (c) limestone and acid phosphate—each on triplicate plots and making chemical analyses of the crops of lespedeza from which the weed crops were separated and both measured. Acid phosphate alone increased the stand of the legume, or conversely, reduced the percentage of weeds in the harvested crop. But phosphate combined with limestone as the soil treatment resulted in no weeds. Percentages of lespedeza in the harvested crop were 60, 90 and 100, respectively, in the order of soil treatments listed above.

The figures for the total harvested phosphorus in the crop in the same order as above were 1.44, 1.78 and 2.53 pounds. The amounts of the harvested calcium, coming along with that increase of phosphorus mobilized from the soil through the presence of the calcium, were 7.12, 8.76 and 13.17 pounds in the crop. The amounts of total protein harvested in the cuttings of lespedeza and weeds combined were 85.2, 100.5 and 182 pounds for the above soil treatments. If the protein in the weeds is subtracted, the values are 73.9, 97.2 and 182.0 pounds delivered by the crop of lespedeza only.

In terms of the concentrations of the phosphorus in the hay, there was the same percentage of it, namely .189% in the dry matter, when the soil was given no treatment as when it was given the combination of lime and phosphate. It was higher, namely .201%, when phosphate only was the soil treatment. The concentrations of calcium in the hay were also higher for the combined treatments of the soil than for the no treatment, but both of these were lower than the concentration of calcium in the lespedeza hay given the phosphate alone as the soil treatment.

All of these data remind us that while we may visualize calcium carbonate uniting with acid phosphate in the soil to make the phosphate less soluble, according as the soil extraction in the laboratory measures the activity of the phosphorus, yet we must visualize some different situations than momentary solution values determining the mobilization of phosphorus from its supply in the soil into the plant root. The presence of calcium in the roots of legumes has demonstrated its significance via the plant's processes in mobilizing the nutrients like phosphorus, nitrogen and potassium from the soil into the crop in other researches. In connection with many more living processes, the phosphorus is closely associated with and improved by the presence of calcium. The same situation prevails in the soil even though, according to usual chemical facts and the tests of the soil, this appears to be a paradox.

LIMITED ACTIVITIES AND SUPPLIES OF SOIL PHOSPHORUS GIVE PLANTS HIDDEN HUNGERS

In the growing of plants, the shortages of phosphorus are not commonly expected to give their manifestations in the seedling or the plant's early growing stages. The seed represents a liberal stored supply. The uptake of phosphorus is also at a relatively low rate as long as the plant is in the vegetative stage. But once the pollen from the anthers drops on the stigma of the pistil to start the process of reproduction, then the uptake of phosphorus by the plant from the soil increases decidedly. This emphasizes the much higher requirements for phosphorus for seed production and for the storage

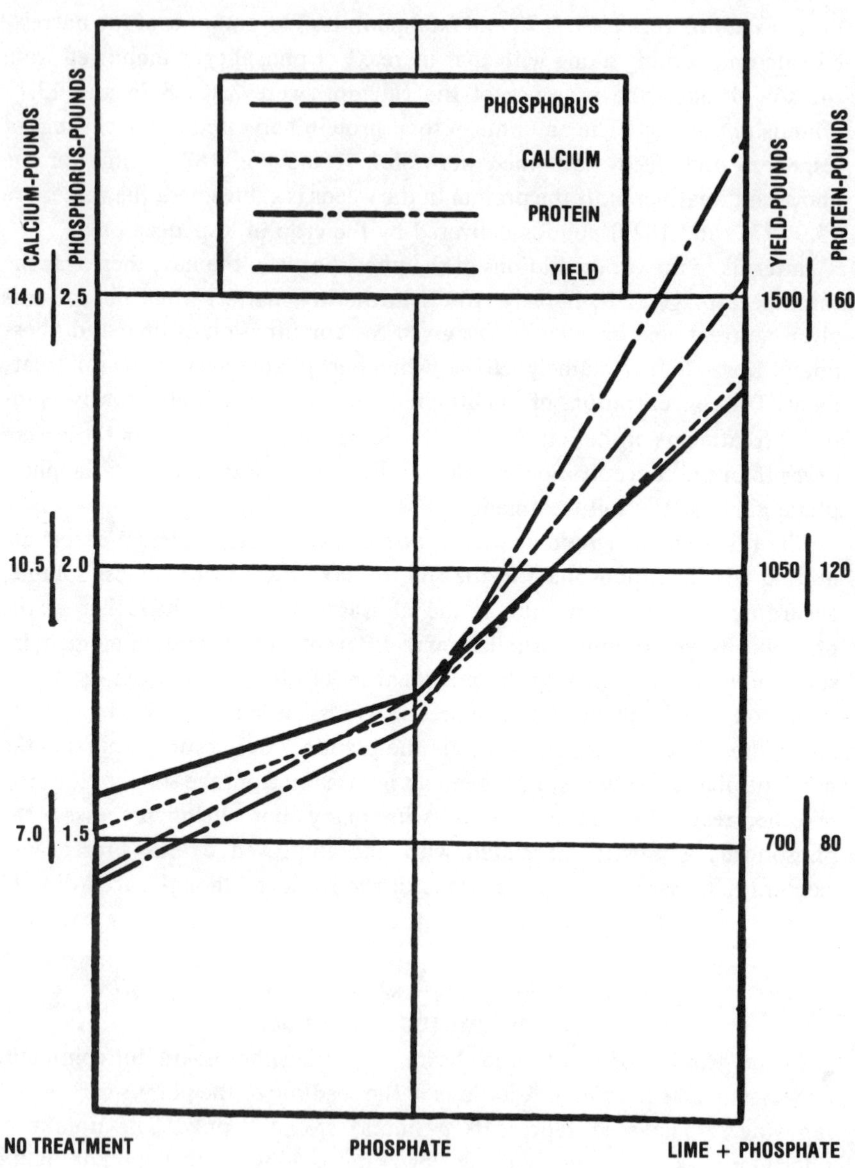

More of the applied phosphorus was mobilized into the lespedeza crop when the soil was also limed. Chemically, limestone makes acid phosphate less active. Biochemically, it made more of it active in moving into the crop.

76 *Soil Fertility and Animal Health*

of phosphorus there, than for the growing of the vegetative mass. It duplicates, to some degree, the cell's slow rate of forming nucleic acid, or sulfur-hydrogen compounds, during the increasing mass of the cell by early growth in contrast to the higher rate of forming these special nucleic compounds apparently required for the division of the cell. This higher uptake of phosphorus results in the .30 to .40% of phosphorus in the wheat grain, for example, while that in the remaining straw amounts to only about .06 to .10% of the dry matter. It illustrates the higher delivery of phosphorus in the grain for the nutrition of the animal in contrast to the problem of animal survival when fed on the straw. Even in the tuber of the potato as the reproductive organ, which may be very low in phosphorus, the relative amount of phosphorus is 3.4% of the tuber ash while it is only 1.2% of the ash of the potato vines. Even this vegetative kind of reproduction illustrated by the potato serves to concentrate the phosphorus to about three times that found in the other vegetative parts of the crop. Root crops exhibit the same phenomena, so that the early statements were commonly made that the roots of plants are high in their concentrations of phosphorus. This is true for the roots of biennials or perennials serving to start the growth of the crop in the next season. Roots of annuals may be only slightly more concentrated in phosphorus than are the vegetative tops. The major differences between concentrations of phosphorus in the roots over those in the tops are in plants of which the root represents reproduction of the next crop.

In our neglect of the fertility of the soil while searching for substitute crops for those failing from this neglect, we have been gradually growing forages offering lowered concentrations of phosphorus. These lowered concentrations in the feed crops are not readily recognized. They are a form of the plant's hidden hunger. The forage plants commonly grown in the midcontinental areas of general farming are not often so low in phosphorus as to represent serious deficiencies. It is commonly considered that when roughages contain near .18% of phosphorus in their dry weight, there is little likelihood of phosphorus deficiency for dairy cows—even when corn is fed as the only carbohydrate. However, there are many areas of underdeveloped and of excessively developed soils put to animal grazing in this and other countries where the vegetation on which the cow is expected to feed herself does not supply enough phosphorus in the kinds of plants growing there naturally.

When the phosphorus in vegetation is very low, the calcium may also be relatively low. The minimum of it in roughages required to prevent its deficiencies for animals has been given as .3% in the dry matter, or at about two and one-half times that of the phosphorus. These values correspond closely with the calcium-phosphorus ratio in the normal blood serum of our animals and of ourselves. Troubles connected with low concentrations of phosphorus in the blood may remain hidden to suggest that the blood samples might profitably be tested for phosphorus when drawn for other purposes. Thereby we might reveal impending troubles coming via the plant growth supposed to be animal nutrition, before serious or disastrous symptoms from that source reveal them. Plant processes of protein production and delivery of a seed crop are threatened with failure when the calcium and phosphorus concentrations drop to such low levels; still not jeopardizing the animal's health with clinical manifestations. Good legume hay, like red clover, carries a concentration of 2.0% nitrogen, 1.4% calcium, .40% magnesium, and .25% phosphorus, suggesting what the pastured animal chooses as its protein supplement and its inorganic element supplement to the grasses in the pastures of mixed herbages. Different plants are merely demonstrating different degrees of protein shortages in quantity and in quality for animal nutrition in relation to the vegetative bulk they produce. This protein deficiency in the plant reflects the shortages and imbalances in the fertility elements among which phosphorus, as a shortage in quantity, and more often in its activity in the soil, is a decided factor.

MANY FACTORS IN THE ANIMAL BODY
CONTROL THE ACTIVITIES OF PHOSPHORUS THERE

Phosphorus deficiencies in the soil may be more specifically connected with hidden hungers of our animals when those troubles can be traced back to what we fed them, and then back to the soils growing the feeds. Phosphorus exhibits its activities always connected with a reserve supply of itself seemingly inactive. This is particularly the case in the animal body. There much of the phosphorus content is hidden away in the bones. But this less active supply there can be mobilized into action rather quickly when a phosphorus-deficient diet for the rat moves the stored phosphorus from the bone into the blood, or the reverse when a high phosphorus diet moves it from the blood into the bone—each within 12 hours after the changes of the required phosphorus supply in the diet promote these shifts. Our failure to know the functions of phosphorus has left us blind to its changing supplies and activities as they indicate our chances to prevent trouble by management of the animal's ration via the fertility ration we are feeding the crop through the soil.

When only about one-half of the blood's content of calcium in the plasma is in the active form, *i.e.*, ionic, while the rest is bound to the protein there, then, when there is the reverse calcium in the bone and when the phosphorus in the body is in almost specific molecular relations and ratios to the calcium, we may well view phosphorus as (a) an inactive reserve in the bone, (b) much

Gross deficiencies of phosphorus in some parts of Texas are called the "stiffs." That is its common name in many other countries. There is no single symptom, just too many of them, as this specimen suggests.

that is inactive in combination with protein in the blood plasma; and (c) a small active supply in the blood stream. But to complicate the activity matter still more, the active phosphorus is represented by three degrees of ionic activity according to its different ionizations. This results from the fact that the phosphate ion is trivalent, or carries three negative electric charges. The phosphate is more active as it is combined with less calcium and more hydrogen or acidity. Here is the means for the body's acid base balance. This is similar to the shifts in acid base balances in the body brought about by enzyme induced alterations of ammonia (a cation) or of biocarbonate (an anion) excretions in the kidney tubule. As another complication in the phosphate activity, there is the fact that phosphate is so closely associated with calcium and this latter, in turn, with carbonic acid in the blood stream and even in the bone, that the activities of phosphorus are modified by the related behaviors of all of these ions. Phosphorus will be properly comprehended, in terms of our keeping it amply supplied in the feeds we raise, only when we know its biochemical activities more completely as well as the amounts of it in the plant ash.

It is significant to note that phosphorus is an inactive combination, apparently, when adsorbed in the protein of the blood plasma. It is also laid down in an inactive form, apparently in the bone. But even there, the basic procedure of bone construction suggests that collagen, the protein part of the bone, is laid down first and then a phosphorus compound, suggesting the mineral apatite, is distributed within that as the matrix for eventual solidification. It is significant to note that in all these activities of phosphorus, they have this element in an organo-phosphatic compound. The activities of phosphorus within the plant emphasize their increase, too, via organic combinations of it. When vitamin D brings about increased activities, it does so through organic combinations. This vitamin serves to raise the activities by increasing the active phosphorus and calcium levels in the blood, through increased absorption of them from the gut, and by improved reabsorption of phosphate from the kidney, to recycle this active form rather than permit the body's excretion of it. The body is thrifty with reference to its supply of active phosphate.

Phosphorus has long been known to be the one essential element required as fertilizer on humid soils almost everywhere if healthy animals are to be raised. Just why it is the more common cause of hidden hunger has not been comprehended. It is, however, the one element of the anionic group which seems to be required to lay the basis for animal health via the protein-rich crops we grow, *i.e.*, much as calcium is the required cationic basis for proteins and all other required ions which usually come along when the soil grows the more complete proteins. Increasing the supply of phosphorus in the soil has long been put up as the companion need going with the need to increase the calcium (magnesium) supply there for healthy animals. The natural climatic pattern developing soils has given the combination of these three in the soil their emphasis in the ecological pattern of wildlife. So while calcium, magnesium and phosphorus are being catalogued in their functions in plants to help us manage the soil more efficiently for healthy animals, we are bidding fair to prevent many hidden hungers. The knowledge so far gained is gratifying enough to encourage us to learn more of how we can manage the soil fertility to grow healthy animals rather than depend on drugs for relief or escape from disaster. We must still grow health into the animals via the soil fertility. It cannot be thrown into them via the chemist's shop. Animals can be creators of their own health through what they eat if we can offer help in that direction by managing the soil treatments accordingly.

CHAPTER 8

Cows are Capable Chemists

IN OUR VARIOUS agricultural attempts to use soil treatments to grow better feed for healthier livestock, we commonly speak of limestone and phosphates as required to make strong bones. We mention those same two as necessary fertilizers to grow legume crops, the particular feeds which are rich in protein as well as in calcium and phosphorus—the major chemical elements composing the animal and human skeletons. We grow legumes as the particular plant which can use the gaseous nitrogen of the soil atmosphere, by the help of the nodule bacteria on their roots and build it into proteins of sufficiently higher quality so they may be supplements to the incomplete proteins in the other hays, the corn and most grains grown extensively on our humid soils. We grow legumes also for their higher concentrations of calcium, phosphorus, magnesium, and other ash elements. We grow them as mineral supplements to the common forages and feeds. We need all the chemistry we can call in to help us build soils to grow nutritious feeds, even with all the purchased protein , vitamin and mineral supplements, just to fatten our castrated males. Even then, the deficiencies and the degenerative body processes are taking such a large toll of their numbers to make us market them as early in their life as possible.

This is the situation, in general, when we are the chemists concocting the feeds which we compel the animal to take while we attempt to feed them most efficiently. For that efficiency, our criterion is the maximum increase in bodyweight for the minimum of feed consumed. The use of that criterion of only weight increase has crowded the life stream of our growing, young animals so badly that the stream is about to be dried up through an increasing crop of dwarfs. These births of the young, too deficient in the capacity to grow and to keep the life stream flowing, have become more common in both beef and dairy cattle, not to emphasize horses. There is a higher percentage of them where the stream of life has been more carefully guided by us according to particular pedigrees. Apparently, as chemists given to feeding these animals with so much economics in our criterion, we are not very able to keep the life streams flowing. We are not as efficient in multiplying the animals in numbers of healthy ones, as we are in fattening the males after surgically eliminating each one's chances to contribute more than itself to a larger life flood.

Perhaps it might be well to learn something about the ways the cow calls the chemistry in compounding her ration for herself, if given the opportunity to demonstrate. Should we set all the different possible constituents of the ration before her, she might make some good suggestions by her choices under different conditions. When it comes to choosing the forages which she grazes, she has always shown a keen chemical sense in her selections of different plant species grown on the same, more highly developed soil. But she shows less discrimination as the soil is less highly developed and consequently more highly fertile. Now that we are using different fertilizer treatments under the same crop and ask her to choose therefrom, she has distinguished more keenly—than any laboratory chemist can—between the same plant species on the same soil given different treatments.

We need to start observing and judging the cow as she is a chemist on the hoof guiding her own nutrition. That observation and the subscription to her suggestions may well be exercised in advance of our judging her merely as so much beef carcass—with more of itself in the higher priced cuts in the pack-

By Courtesy W. M. Landess, TVA

The cow is not merely a mower of grass. She is a capable connoisseur of the vegetation as it satisfies her physiological requirements for growth, protection against disease, and reproduction. She seems to balance her diet well if mixed herbage permits.

ing house refrigerator or in the cellophane wrapper in the counters of the store's meat display. Cows must have always been chemists of renowned capabilities to have done so well in keeping the stream of their own lives flowing all these years in spite of us, rather than because of us.

AS A CHEMIST BY EXPERIENCE AND SURVIVAL, NOT BY ACADEMIC TRAINING, THE COW LED THE NOMAD OVER FERTILE SOILS

One needs only to look at history to realize that the survival of the nomad in his wanderings and a primitive agriculture depended on the fact that the cow went ahead of the plow. She was truly leading the people and really decided the geographic direction in which they and their agriculture went. She inspected the natural forages, she delineated the areas of fertile soils, and she labeled them as fitted to grow food for her owner as well as feed for herself. The agriculture of the Old World followed with the plow where the cow had first gone to recommend that this implement should be put. Such an agriculture has long endured because it sent this capable nutrition chemist ahead to scout the areas for all that meets her requirements.

Our American agriculture developed by quite the reverse of that procedure. On much of our arable land area the plow went ahead of the cow. We used no such capable chemist, like the nomad had, to put the stamp of approval on the fertility of the soil as a suitable and enduring food creator for both the cows and ourselves. We are now coming gradually to see that (a) in our problem of protein feed supplements, (b) in the increasing so-called disease, (c) in the irregularities in concepts by the cows (and other livestock) and (d) in their failures in reproducing. There are powerful suggestions that the soils may be deficient in items which we have not yet catalogued for their nutritional significance. Perhaps by organizing a research team of different individual talents we may profit by including the cow in the crowd aiming to contribute to the biochemistry as well as to the inorganic chemistry of soil fertility and plant nutrition for better animal and human nutrition.

THE COW RECOGNIZES UNBALANCED FERTILITY MARKING HER DROPPINGS BY EXTRA GREEN GRASS

It has taken us a long time to learn that while a plant is making vegetative bulk, such a performance is not necessarily proof that it is making feed. The cow has always been showing her recognition of this significant truth. She has regularly refused to take the tall, green grass outlining the spot of her droppings. Yet she was eating shorter the short grass around it. We believed her merely fastidious about getting in contact with her own voidings, both urine and feces. But after setting up some feeding trials with rabbits in a so-called scientific research project, we got the suggestion clearly that she was merely telling us that too much nitrogen as a fertilizer under imbalance to other elements makes a grass which she as a chemist and nutritionist cannot approve for her own consumption. After that experimental expenditure in money, time and study, it became quite evident what the cow was telling us. She was corroborating what the livery horses told their New England owners who refused to buy timothy hay fertilized with only Chile saltpeter, or sodium nitrate.

In using the experimental rabbits to confirm the cow's testimony about unbalanced fertilizers as the cause of unbalanced nutrition in her feeds, increasing amounts of nitrogen were applied as top-dressings on a pasture with a mixed grass-legume flora. When only more nitrogen was applied and the resulting tall, luscious, green grass was made into hay for feeding trials with the rabbits, they were increasingly reluctant to take it as more nitrogen fertilizer was used, except under an approach to starvation. Their loss of weight and their increased discard of hay from the feedrack—to use it only for bedding—were ample suggestions from the rabbits' behavior that the report from the art of agriculture via the cow's refusal was just as telling about unbalanced soil treatment, giving unbalanced nutrition in the forages grown

The spots of taller and greener grass in the pasture report the cow's chemical sense to avoid the excess of nitrogen applied in her feces and urine. The latter adds more than the former to make the urine spots with taller grass than the feces spots. The cow disregards the greener grass to let that attract our attention but not hers.

thereby, as was the report by the science of agriculture via the test rabbits
refusal. The rabbits as chemists merely gave the same vote as the cow on the
simple matter of balanced fertility being a requirement for growing balanced
nutrition.

Too much nitrogen is what both the cow as farm livestock and the rabbit as
a living kind of scientific laboratory equipment were saying. The one was
referring to the cow's urine dropped on the spot in the pasture and the other
was referring to commercially fabricated fertilizer nitrogen. Both were
capable as biochemists and as chemists. But with our attention fixed on
quantity yield, we had not recognized the simple fact that they were making
an unfavorable report on the *quality* of the feeds according to the *unbalanced
fertility* of the soil growing them. Here was a case where we as researchers—who are merely trying to learn more—discovered that the cow was a better soil chemist and biochemist than we are.

**THE COW'S DISCRIMINATING GRAZING ACCORDING TO SOIL TREATMENTS
IS A WISE CHECK ON TEST TUBE RECOMMENDATIONS OF THEM**

Though we may have put the plow ahead of the cow, nevertheless she is
following along behind it to keep reporting on how successfully we are using
soil treatments to grow feeds satisfactory to her. We need to make certain,
however, that her reports are accepted as suggestions to be followed for preventions and not as postmortems. Apropos thereto, her selections of certain
pasture areas given different fertilizer applications are coming to be the
guide for many farmers as to how their soils should be treated and managed.
Farmers of Missouri are using not only the laboratory test tubes on the soil to
direct their soil management of fertilizers, but are also asking for the cow's
assay of the forages according to her preferred choice of some fertilized areas
over others. They are respecting her capabilities as a nutrition chemist according to her biological criterion of undergirding her own nutrition rather
than accepting our economic criterion of cheap gains in weight. They have
invested in their own soil-testing laboratories in already almost 90% of the
counties of the what is said to be a livestock state. They are doing this
because they see the better soil treatments reflected in better health of their
animals.

Lime has long been used on Missouri soils because the cow recommended
it by her choice of limed areas as preferred grazing over those unlimed. She
exhibited such choice where liming as an applied carbonate did not modify
the pH of the soil or change its degree of acidity one iota. But chemical tests
of the improved forage quality indicated that by her choice, she suggested
that liming was a process of fertilizing with calcium to improve the plant's
nutrition for the synthesis by it of more protein, for the delivery of more con-

The stakes marking out the line limiting the soil treatment (right) served also, apparently, to limit the shorter grazing of the grass by the cattle. They let the "weeds" in the thinner, unfertilized grass stand (left) grow taller and "take the grass." Cows are capable chemists inspecting the crop for its values as their nutrition.

centrated ash elements, and even, possibly, for higher concentrations of sugar. She has even chosen cornstalks—after the picking of the corn—on the limed portion in preference to those on the unlimed portion of the cornfield.

Now that lime has been widely used and potassium is a more recent addition to the soil, the sections of the field given potassium are getting the cow's first choice to tell us of the soil's needs for more than only calcium or lime. She has also been grazing the phosphated soils as first choice, regardless of whether the rock phosphate was used as a sustaining mineral fertility or the superphosphate served as a starter fertilizer. Since the latter adds calcium, magnesium, sulfur and a host of trace elements—made more active or available by the sulfuric acid treatment of the original rock phosphate—we have thought that she was reporting her choice of grazing in relation to only the phosphorus treatment of the soil. She may have been responding to the effects, also, of the trace elements and other neglected ones, like magnesium and sulfur so common in superphosphate, which we ourselves and as her feeders do not yet consider as widespread deficiencies in the feed, much less in the soils growing it. The cow may well be guiding our fertilizer practices to the refined degree of including the trace elements. She may well be in the bioassayer in the field to supplement the wisdom collected in the laboratory. She

When the rounded blank corners, resulting from drilling the barley, were drilled in to give doubled fertilizer application on turning around in this 40 acre field, the steers put on for fall grazing took to these doubly fertilized areas first, or spots where extra fertilizer was spilled to suggest a heavier application than was first used after the soil was tested.

is casting doubt on our use of nitrogen fertilizers too freely and has always exhibited her doubt in that respect. Apparently we must experience some disasters and hold postmortems on both crops and cows to learn when there is too much nitrogen used on the soil, or not enough of all else.

HOG'S CHOICES SUGGEST THAT THE COW, TOO, IS MORE
OF AN ORGANIC THAN AN INORGANIC CHEMIST

For us to believe that the cow is a chemist capable enough to be pointing out the values of single inorganic elements, as we do in burning or ashing the forages and making analyses in the laboratory for the separate chemical elements, may be crediting her through an excessive stretch of our own imagination. She is very probably indicating the presence not of the separate inorganic or mineral fractions, but rather of organic compounds synthesized in the plant by the help of these inorganic agencies coming from the soil in certain relative amounts.

That these are the probable facts is suggested by some tests of choices by hogs in which they selected the different corn grains from the different compartments of the self-feeder, which represented plots with different soil treatments where those grains were grown. The plots in each of two series were given increasing combinations of different fertilizers and grown to sweet clover in the wheat ahead of the corn to follow as the next grain crop. In one of these series the sweet clover was plowed under as a green manure crop in early May of its second year, just ahead of the corn planting. In the duplicate fertilizer-plot series, the second-year sweet clover was grown to a seed crop, the residues were allowed to accumulate, and the land was then plowed with these turned under ahead of the corn.

Choices by the hogs of the different corn grains from the different compartments of the self-feeder, representing the respective soil treatments, gave the highest percentage consumption according to the higher amounts of fertilizer used when the dried residues of the sweet clover crop were the organic matter turned under ahead of the corn crop. But the order of choices of grains by the hogs was exactly the reverse when the residues turned under ahead of the corn crop were the green sweet clover as a green manure. Here was the suggestion that the grain consumed is not chosen in direct relation to

Hogs took out the corn first in the corner of the field where the soil was limed some years before and given phosphate to grow alfalfa. They were choosing the grain, not the forage, so precisely that they went to the most distant corner of the 40 acre field from the water and supplement to show their keen discrimination.

the inorganic or mineral elements of the fertilizer put on the soil ahead of the corn grown for this experimental test. But apparently the hogs exercised their choice according to the different organic compounds in the sweet clover reflected in the corn grain. They voted against those from this crop as a green manure, and possibly its resulting microbial byproducts taken up by the corn plant probably directly. But they voted for those in the matured, dried sweet clover residues plowed into the soil in advance of the planting of the corn crop.

When the hog, by its choice of the corn grain, votes down the use of sweet clover in a green manure as the nitrogenous fertilizer for that corn, it is apparently giving the same party its support by ballot in which the cow votes when turned in to graze the green sweet clover in the spring but refuses to eat this bitter legume until she has cleaned up all the other herbage along fence rows and water courses and is compelled, by threat of starvation, to nibble the growing points of the sweet clover. Is it possible that the cow has a systematic allergy to the dangerous dicumarol, the anticoagulant produced by this nonacceptable legume? Has Mother Nature given her an instinct of protection against this organic compound is sweet clover which keeps her blood from clotting so that she would bleed to death quickly on injury, which fact some folks learned through their killing of many head by dehorning them when they had been previously compelled to eat sweet clover over hay?

Apparently the cow, in her selection of different herbages, is acting like a biochemist searching out the organic food compounds and is not only an inorganic chemist ashing them to determine the inorganic or mineral contents. Perhaps when so much of her ration commonly consists of the bulky carbohydrates or energy foods, she is searching for the proteins as special organic essentials to balance them. Since the proteins are usually synthesized more bountifully in relation to carbohydrates by forages requiring inorganic fertilizers generously, like lime, phosphate and trace elements on our well-developed or humid soils, we may be erroneously crediting her with too much as an inorganic chemist searching for, and detecting, only the so-called minerals in the feed's ash. In reality, she should be more correctly credited as an organic chemist. Or perhaps we should even credit her as a biochemist struggling to balance her diet according as the different soils with higher levels of fertility have elaborated more proteins, vitamins, etc. in the different plants to make this balancing effort successful within the range of territory she can cover. Are we not sadly mistaken about her abilities to rate her ration as good nutrition when we try to help her out by offering her ground limestone, pulverized phosphate rock and others as inorganic compounds in the mineral box for direct consumption rather than offering her their effects on feed plants via the soil?

AS SYNTHETIC CHEMIST THE COW COMPOUNDS COBALT INTO VITAMIN B_{12}
AS AN INVITATION TO PIGS AND CHICKENS TO FOLLOW HER FOR HER DROPPINGS

Since the cow is one of the higher forms, we may well expect a higher number of elements (both major and minor or trace) and compounds to be required in her feed and in the more complex physiological processes which support her. Equipped as she is to range over extensive territory, she increases her chances to gather from greater soil areas all the requisites from that source. Then also when she takes daily about 150 pounds of green feed into her paunch for microbial as well as alimentary digestion, she is *synthesizing* as well as *analyzing* many compounds. That synthetic performance—a symbiosis between microbes and herself and vice versa—is the means whereby she can be fed chemical nitrogen in urea form and can synthesize it into protein-like compounds, via the microbes, to be digested and absorbed into her bloodstream farther along in transit through her digestive canal. It is that means whereby she synthesizes several vitamins, hormones, etc., for her own benefit; seemingly recognized long ago by the hogs and the chickens following her but recognized by the nutritionists only recently. Through this recognition, the dried cow dung became a recommended ingredient in the poultry ration on behalf of some factors only partially known, including now vitamin B_{12}, the cobalt vitamin.

Surely the cow that is a collector of all the requisites for these microbial synthetic processes which are serving her, and passing some benefits on to the pigs and chickens following her for her droppings, must be a capable chemist of wide scope in her own right. She was this long before we even imagined those highly integrated nutritional relations of different life forms, including the interdependencies between microbes, plants, pigs, chickens and cows. The art of agriculture has long had the farmer's pigs following the fattening steers. This was a practice in that art for decades before the relations were recognized by the science of agriculture which prescribed dried cow manure in the rations for chickens, but not for pigs, before the vitamins in question were established or commercially available. Surely the venerable art of agriculture has been following the lead of the cow long before the youthful science which is gradually coming along to explain.

The cow as a ruminant has not been much appreciated as a synthetic chemist either in the organic or the inorganic areas. We have not asked ourselves whether she isn't better equipped to synthesize many organic essentials, as well as those combined with the inorganic trace elements like vitamin B_{12}, to make more enzymes and carry out reactions thereby than most any other animal? Is this capability of hers in the big chemical realm the reason she can be strictly herbivorous while hogs and chickens are more carnivorous, *i.e.*, must be fed some animal protein supplement? Did she know that some of the trace elements used as fertilizers encourage higher concentrations of certain amino acids in crops, like alfalfa, to make it more nearly a complete protein as has been reported by the Missouri Experiment Station? Apparently she has been familiar with the fact that such soil treatments and high protein quality of her feeds serve to give a health that guards against so-called contagious abortion. To a milk producer like the cow, the vitamin B_{12}, considered an animal-protein factor and a lactobacillus factor, must have long been a familiar subject. It is not too much of a stretch of the imagination to consider manganese, zinc, copper and other trace elements playing similarly significant roles in being parts of certain vitamins, enzymes, hormones and other biochemical tools in the synthetic performances, still unknown, which the cow has been carrying out. Perhaps the cow has been more of an analyzing and synthesizing chemist of the inorganic and organic compounds than we recognize.

HAVE WE MADE THE COW A VICTIM OF HER SWEET TOOTH?

Whether the cow searches out her carbohydrates as discriminatingly as she does her proteins has been doubted when it was said that cows are fooled by sweetness in their taste. It was told that ordinary winter-dry broom sedge (*Andropogon virginicus*) sprayed with a solution of saccharine, to provide the sweet taste only but no sugar, will be eaten readily by cows. This has been considered as evidence that the cow can be fooled by her sweet tooth and that she has no sense of her nutritional requirements so far as carbohydrates are concerned. This would imply that she is not a carbohydrate chemist.

This may have some validity if we see it as a case of her being starved to the point of having fallen victim to a sweet tooth, or a readiness to fill with anything of as little value as the broom sedge, which she commonly leaves untouched, to be the common winter cover on many an infertile pasture. Yet when we consider her suffering with bloat, the studies on the sugar content of legume forages, in relation to soil fertility, give theoretical grounds for considering that affliction the result of ingesting sugars in excessive concentrations and amounts to bring on an indigestion interrupting the parastalsis that lifts the front part of the paunch. Imbalanced fertility growing soybean forage demonstrated clearly the failure of the sugars to be converted into protein and their accumulation to a high percentage in the dry matter. Might it be possible that we have so limited the cow's grazing and so exhausted the soil fertility that when under severe hunger and nothing but high-sugar legumes, she gets an indigestion resulting in a fermentation and bloat as the symptom more than the only cause of this common trouble?

Some recent reports with rats tell us that while they take to the sweet taste of saccharine on just about a par with the sweet taste of dextrose when on maintenance diet, yet when put to exercise, like swimming, they demonstrate decidedly that they will take much more dextrose as the energy source and do not take to the saccharine of only the sweet taste.

The rats and their behavior as dietary chemists would lead us to put the cow as a chemist on an equal ability. We would question whether the cow has fallen a victim of a sweet tooth were we not possibly responsible for the perversion of her taste under our limiting the conditions within which she can exercise her discriminating talents as a chemist. In the wild, or on the range, it would seem well to believe the cow a capable carbohydrate chemist too.

FARMER OBSERVATIONS TELL US THAT COWS KNOW THEIR HAY ACCORDING TO THE SOILS GROWING IT

As a chemist, the cow can scarcely be expected to categorize her discrimination according to inorganic or organic compounds, or according to the carbohydrates or the proteins or even as to any particular vitamin, as we might do in our chemical refinement. Nevertheless her keen discriminations suggest a refinement that still challenges that of even any late procedures for accuracy according to observations reported by some farmers.

As a case in illustration, the cows on the farm of Eugene M. Poirot, Golden City, Missouri selected one of four haystacks according to the particular fertilizer treatment of 1936 which lasted in its effects for eight years, or through 1943, but failed to tempt the cattle during the ninth year, 1944. It was not until this ninth year that the cattle no longer demonstrated their annual discrimination between the one stack containing some hay from fertilized soil and the other three stacks with hay entirely from soils given no treatment.

**Food Selections by Ten Rats on Modified Self-Selection Diets
Prior to and During Swimming**

Food Component	Average Intake During	
	Rest	Swimming
Dextrose	3.69 cc	43.00 cc
Saccharine	3.46 cc	2.40 cc
Water	20.53 cc	6.71 cc
McCollum Mixture	12.74 gm	5.02 gm

From Griffiths, W.J., Jr., and Gallagher, T.J., Science 118:780.

It was the spring of 1936—the well remembered year of the severe drought in Missouri—when less than five acres at one end of a 100 acre virgin prairie were used as plots with fertilizers drilled on the surface— each at rates of 300 pounds per acre. In order to provide calcium, the nitrogen was applied in the form of calcium cyanamid. Then the next plot treatment was superphosphate and the third one was these two combined—amounting to a total fertilizer rate of 600 pounds per acre. The total area of the treatments amounted to less than five, or more nearly four acres.

After a study during the summer of the effects of these treatments on the flora, the yield, and on the chemical composition of the vegetation, which gave no significant chemical or other differences, the grass was cut for hay and put up as four stacks—each containing the hay from 25 acres. The hay from the treated five acres was mixed with that from about four times as many untreated acres in making the first stack at the distant end of the field, relative to shelter, salt and water. Three additional stacks in the rest of the field represented the hay from the remaining area of soil given no treatments. This field of four stacks had long served before and since 1936 as the winter feed with the cattle turned in to eat from the stacks. No soil treatment has followed that single one of 1936.

The difference in quality of hay in the stack of 25 acres of it, when only that from five acres of those 25 had been fertilized, was reported by the cattle in the first early winter, 1936. They went to this stack at the distant end of the prairie area daily from shelter and water in disregard of the three nearer stacks. They consumed this far one first with the small part of its hay from the fertilized soil. The entire stack was taken before the other three were.

For every year from 1936 to 1943 inclusive, the grass was similarly harvested and handled as hay with no fertilization subsequent to that of 1936. Each year during that period the cattle duplicated their original demonstration. They chose among the four stacks, and consumed first the one of which a part of the hay was grown on soil fertilized but once.

The year of 1943 demonstrated nicely the fact that the difference in quality of the hay as a result of one fertilizer treatment lasted for at least eight crops or through 1943. This was a season of ample summer rainfall. While making the hay, that from the once-fertilized five acres was put in first, as was the custom. The stack as started, however, was found too small to let the stacking machine put in all the hay from 25 acres. Consequently, the stack was extended by putting the hay from the unfertilized soil from the balance of the 25 acres into an added end of it. When all this hay mixture containing that from the fertilized soil in the initially made part of the stack was taken by the cattle, there was this end of the stack left standing. The cattle then distributed themselves about the other three stacks as readily as about this remnant one. The discrimination had been still very sharp for the fertilized hay that was the eighth crop following the soil treatment and what was only one-fifth of the total hay in the stack. Does such a demonstration leave any doubt about the cows as capable chemists in utmost refinement of their accuracy in detail?

During the late fall of 1944, when there was considerable grazing left because of the late date of the killing frost and when the cattle were turned into this field to start them on the consumption of the stacks of the winter's hay supply, it was significant to observe that the cattle grazed first that part of the field on which fertilizer was applied in 1936. This animal manifestation suggested that the better quality of the feed was still recognizable in the green grass and that the cattle as chemists were still recording the small effects remaining and extending into the ninth crop.

However, later in the winter season of 1944 after the cattle took to the hay, they no longer showed discrimination between the different haystacks. They were consuming the nearby, unfertilized haystack as readily as the distant one containing the hay from soil fertilized in 1936 and mixed with much grown on unfertilized soil. The cows were evidently not capable enough as chemists to distinguish the nutritional effects in the hay from the soil treatments lasting as long as 1944, even though they were still doing so for the green grass at the extended date.

COWS AS CHEMISTS PREMISE THE PROSPECTS FOR LIVESTOCK
ON MORE STUDY OF SOIL CHEMISTRY IN RELATION
TO BETTER PLANT NUTRITION AND BETTER FEEDS

That cows are capable chemists in all they can do for their own preservation against degeneration in health and for their reproduction by their choices of nutrition as the different fertility of the soil grows it is a concept of decided value in keeping our cows in better health. Cows judge the differences in the crop on different soils so well that they need no fences to exclude them from the crops not worth their time to consume them. A single wire needs little electric charge to retain them within the area where fertile soils are growing nutritious forages. Cows are not mimicking fertilizer chemists by inspecting fertilizers to check on the nitrogen, phosphorus and potassium contents put on the soil in water soluble form. Rather they are inspecting the forage crops (we offer them) for their values as nutrition, health and reproductive potential. They are reporting on these according as our maintenance and building of soil fertility are serving the crops with their nutrition in balance for carbohydrate delivery in balance with proteins, plentiful and complete in the necessary amino acids. Cows are chemists without aiming at fattening a carcass. They are aiming to accomplish what we see as the functions of proteins in the body nourished also with all other essentials, namely, (a) the multiplication of cells in the growth of the body, (b) the protection of themselves against being digested by foreign proteins, and (c) the continued reproduction of the species to keep the stream of its life flowing.

There are ample suggestions from the cows, as the progenitors of the life stream of their species, challenging any of us as soil managers to grow crops according to a newer criterion our cows offer. That criterion calls for soil treatments to grow crops delivering to the cow the carbohydrates balanced with proteins complete enough, at least, to keep the reproductive part of the livestock stream flowing without the degenerations labeled "disease" and those giving birth of midgets devoid of the capacity to grow. The cow as a soil chemist is giving us, as chemists of similar claim, the challenge to build the soil fertility as the real creative foundation of life if all the life of agriculture and the rest of us are to survive. The prospects for the future of livestock, if carefully surveyed with the help of the cows as capable agricultural chemists, will be higher only as we view them in terms of the soil fertility as it grows healthy livestock via better nutrition.

BRUCELLOSIS

If you want to reduce human medicine or veterinary medicine to a common denominator, you have to remember that when the animal's physiology is deranged, it doesn't make much difference what you call the problem—but it is very probably a mistake in nutrition often founded on the attempt to be economical. I have come to the conclusion that deficiencies are more often at the base of health irregularities than we realize. I have had occasion to test some trace elements. And two M.D.s hooked up with me at one point. So I jumped in at the deep end because if I got into trouble with the medical profession—and that's very easy to do—I'd have two M.D.s to pull me out. I put a student on studying brucellosis contagious abortion, which they call *contagion,* which it isn't at all. And we proved it with four generations of a herd of 85 milk cows that were labeled to be slaughtered. We fed them trace elements and we treated the soil with trace elements while we were getting ready to feed the animals the products of the soil. In four years we had 17 female calves that became heifers and raised calves, and their calves were clean according to the veterinary tests. Because you see they introduced an artificial microbe they call a *Brucellosis abortus* as though it were a grand name. It is nothing but the symptom name given to the microbe. We had 17 heifers mature after we started feeding trace elements, and they gave us calves, and those 17 calves were as clean as could be by any test the veterinarian could run on the bloodstream.—*Dr. William A. Albrecht, in an interview with Acres U.S.A.*

CHAPTER 9

Crops Must Grow Their Own Soil Organic Matter

CROP ROTATIONS, or a sequence of different crops on the same soil, have had pronounced emphasis in our soil and crop management. There has been this emphasis in spite of the fact that nature grows her crops to the climax of each of them by keeping the same crop continuously on the same soil and in the same place. Crop rotations are the most economical way of harvesting maximum crops. Consequently, they are also the quickest way of depleting the soil most rapidly of its inorganic fertility under such crop removal.

The records of critical studies of extended cropping systems and of changing soil fertility supplies have established this fact clearly. In soil management, we use limestone, rock phosphate and processed inorganic salts as fertilizers to put back some of the inorganic elements which plants use and crops remove. But in that practice, we are not reminding ourselves that nature is doing that too in building to a climax of crop production. But she is also putting back organic fertilizers simultaneously. Her return of fertility in both the inorganic and the organic forms is equal to the contents of the entire crop grown. We have been neglecting this latter half, namely the organic portion, of the requirements of managing soils as well as nature has been managing them before we took them over under our management.

Experiences are accumulating which show healthier plants, less pests and higher quality of crops when more organic matter and less chemical salts represent the return of fertility in the efforts to rebuild and to maintain that phase of the soil. The whole subject of healthy microbes, healthy plants and healthy animals needs to be studied in our agriculture when that complete return of all fertility elements in their organic combinations serves in nature's way of growing the crops of all life forms to a climax, without ministrations of drugs, poisons and medicines.

It might be helpful to observe that natural crops, or those in the wild, exhibit their growth year after year in the same place. When any crop finds the inorganic fertility in proper balance for its specific physiological processes, then that crop takes over. It becomes bigger and better annually until it reaches what is called an ecological climax. Then it declines gradually. This is a case of a particular plant fitting into all the factors of growth, self-protection and reproduction in that particular location better than any other plant would fit. Grasses observed in the wild suggest that the inorganic fertility must be more delicately and more properly balanced for each particular species in one place than in another. On the high mountains where disintegrating rocks are washed into a lower area, we marvel at the pure stand of some flowering species. The botanist observes and inquires, "That is an ecological climax. Why can't we manage our crops to have them equally as healthy and grow equally as well?" He too forgets the contribution by the organic matter as essential fertility itself. We have seen the ash chemistry but not the organic chemistry in the soil under nutrition of healthy plants.

The natural forests are similar illustrations. In starting a planting of one for conservation's sake, the forester says, "We must establish a pure stand and build up its forest floor which excludes all else." He means to tell us that when the soil of the forest is well covered with the dead leaves or needles and other forms of organic matter contributed by that particular species, then the forest trees will grow best and be healthiest. They will grow toward their ecological climax, namely a pure stand of them with no defective ones, each capable of protecting itself against diseases and pests and exhibiting the ecological law of unique fitness for that soil-climatic setting to the exclusion of other species.

In the climaxes of nature, there is always illustrated the proper setting so far as the cycle of accumulated inorganic soil fertility (with small annual additions from the mineral part of the soil) feeds that crop more completely for survival there than it would any other crop. In agronomic research, there is now the increasing effort to learn how to meet each crop's need for inorganic fertilizers in proper ratios and amounts by growing it continuously rather than in a rotation. We are, thereby, learning slowly how different crops vary in what they create in their organic products as animal nutrition. We are learning also how those products from the same crop vary because the fertility of the soil nourishing it varies.

We have done little, however, to learn what that creative procedure does to keep each crop healthy in that setting of complete return of the crop as the organic fertilizer for the next crop. By that emphasis on only the inorganic nutrition, and by growing the plants in one place continuously, we have succeeded in producing much yield as crop bulk. At the same time there has too often resulted, (a) lowered nutritional value of it as feed, (b) plants less healthy in terms of their own internal protection against microbial and insect attacks and (c) a failing reproduction or a decline of the species. These facts are a powerful suggestion that we have failed to consider the simple truth that in nature's ecological climax, not only the simple organic elements in

This valley in the Santa Rita Mountains was an ecological climax of nutritious grass in 1903 (upper photo). It became such, in man's absence, by the return of the organic matter of the crop annually to be the major fertility supply for the next crop. Cattle taking off the grass robbed that climax. Forty years later (lower photo), there was so little return of organic fertility and so little mineral breakdown that only scattered trees of the leguminous mesquite could survive, but no cattle.

cycle of removal from, and complete return to, the soil are serving in plant nutrition, but also that plants are nourished but taking up, as well, in their nutrition many organic compounds which are in the same cycle. There is the further evidence that organic compounds resulting from the decay of past generations of their own particular kind serve as their organic nutrition to a greater advantage than dead generations of other plant species. Such facts tell us that in the struggle to grow healthy life forms rather than sick ones in agricultural production, we ought to turn our attention to the organic matter which crops produce in themselves and return to the soil as it is organic nutrition for the next crop. The cow's health ought to profit by such attention as much as anything else that agriculture produces.

ROTATIONS, THE QUICKEST WAY TO MINE THE SOIL

How the belief and the implicit faith in crop rotations as a help in maintaining the soil's productive capacity became so well established is difficult to understand. It is self-evident from industry that one cannot produce and ship out a manufactured product regularly without bringing in raw materials continually. Surely agricultural production via soil as the factory is no exception. Yet in an extensive reference work published no later than 1954, there is the following statement under the heading of *Maintenance of Soil Productivity*. The writer says, "Improved crop rotations are of greater importance than any single practice . . . It would be impossible to over-emphasize the importance of crop rotations to control diseases, maintain fertility, prevent erosion and maintain soil structure. Each farmer and gardener must learn which rotations are best for his soil."

This contention implies that the mere rotation, or a kind of crop juggling, does good, *per se*, for the soil. According to the results from long continued cropping on many experimental station fields, quite the opposite is the truth. Rotations are the speediest way to exhaust the soil's productivity. They do not maintain the fertility of the farmer's soil. He himself must do that in terms of both its inorganic and organic matters returned for their maintenance as the supplies that are built into crops.

Studies at the Missouri Agricultural Experiment Station on Sanborn Field, one of the several oldest in the United States, contradict the report of the above quotation. Since 1888 some rotations of two, three, four and six years in length, and of various combinations of corn, oats, wheat, clover and timothy have had their modifying effects on the maintenance of the soil fertility under critical chemical measures regularly. The longest rotation, namely six years of corn, oats, wheat, clover, timothy and timothy—for which sod part of three years much conservation effect is usually claimed—has exhausted the soil of its fertility to a lower level than even the soil under continuous cropping by corn, oats, wheat and timothy for the several fertility elements commonly considered. The soil of the plot under the six-year rotation, with all crops removed and nothing but seeding returned, was the lowest of nearly forty plots in the supply of exchangeable potassium. Of all the untreated plots, it was the lowest in exchangeable magnesium. In calcium, its exchangeable supply represented only 30% of saturation, when 75% is considered the optimum. Its pH value was 4.6, one of the lowest and thereby one of the most acid plots of this old field. In phosphorus, it was next to the lowest of all the plots, namely, the one in continuous timothy with no soil treatment, and both of such a low value for phosphorus for which even the test itself cannot be highly discriminatory. Its supply of organic matter was also next to the lowest of all plots, namely, the soil under continuous corn with no soil treatment since 1888 and all crops removed.

Since the differences in the organic matter can be measured by the differences in the total nitrogen of the soil, it is significant to note that the six year rotation during 50 years of cropping, with crop removal and no soil treatment, depleted the soil's supply of this phase of fertility more than was true for the separate crops in the rotation grown continuously. Where the total loss of nitrogen from the soil's original supply under the rotated crops was 1180 pounds per acre, the average loss of nitrogen from the several plots growing each of the same crops continuously was only 1070 pounds per acre. In terms of this loss from the soil's original stock of fertility, namely the burning out of the virgin organic matter, the longest rotation mined the soil faster than the average of the separate continuous crops.

We dare no longer say, if we are to be truthful to the facts of nature, that "Improved crop rotations are of greater importance than any single practice," when speaking of the maintenance of soil productivity. Rotations do not live up to such claims for them when a six year rotation in 50 years was so destructive to what may well be called "the constitution of the soil." Even this long rotation in its late years with nearly three years of no crop removal because of crop failures of clover, timothy and timothy, was also near failures of the crops of corn, oats and wheat in spite of their following those three years which were almost a fallow treatment for these grain crops. Here is evidence, like that from ecological climaxes, suggesting that if we must start with exhausted soils, the problem may be less difficult and complex in building back and maintaining the fertility if we use one crop and grow it continuously. That procedure would seem a logical one when the evidence shows that rotations were the quickest way of mining the soil by calling in several different crops in rapid sequence, each for its different and added exploitive effects. We shall not learn how to maintain soil fertility under a kind of crop juggling as quickly as under continuous cropping. That knowledge dare not arrive much later.

SOIL WITH MORE ORGANIC MATTER GROWS SEEDS OF HIGHER QUALITY

That the ecological climax, by its return of the organic matter from each succeeding dead generation, was helpful in the survival of the plant species is suggested by the lower quality of the seed wheat when grown continuously with nothing returned, than when organic matter in six tons of barnyard manure was returned with wheat grown continuously and the entire crop removed. Tests of the seedling vigor of grains from these plots by Dr. R. L. Fox reported that of the wheat seeds grown with no soil treatment only 42% showed emergence of seedlings. But where organic matter as barnyard manure had been going back annually, 75% of the seeds had their seedlings emerge to represent that high degree of survival of the species in the next crop.

That such improved effect on seed as potential survival was not due to the nitrogen in the manure growing the seed is suggested by some additional study by Dr. Fox. Urea nitrogen, put into the wheat plant by spraying it on to make the plant much greener during its later growing stage, did not give as good a quality of seed when measured by seedling vigor and early emergence as when the fertilizer nitrogen was put into the soil either with, or in advance of, the seedling. Nitrogen moving into the plant during its later growth stages decreased the seedling vigor. Such late arrival into the plant serves to deposit nitrogen into the endosperm in the forms of the simpler and more common amino acids of the proteins. It seems to come along too late to get into the embryo or the reproductive part of the seed. It is, therefore, of no significance in starting more embryos by which more seeds and more bushels per acre are brought about. But when additional phosphorus is present in the soil, these effects by small amounts of nitrogen are improved through the mobilization of the phosphorus into the wheat grain more generously in company with the nitrogen. The phosphorus in the embryos showing differing seedling vigor varies more than it does in the endosperm. This is the reverse of the behavior of the nitrogen.

As testimony to the effects on seed vigor by the fertility combined in the form of organic fertilizer, there is an additional report from these studies. Wheat growing continuously with only five bushels as the seed yield gave a survival potential, when measured under germination test by percentage emergence, which was higher than that of the wheat seeds from much larger acre yields grown on soils heavily treated with concentrated salt fertilizers. Here are facts which emphasize the contributions by soil organic matter, scant though the supply may be when contributed by the crop itself. They suggest also that the physiological processes are not simple by which the plant creates its own component reproductive organic matters, like the embryo and the endosperm, through which the future plant generations are procreated and the species survive.

THE GROWING SEASON NEED ALSO BE THE DECAYING SEASON

When nature grows a crop in man's absence, she uses a fertilizing procedure quite different from ours when we apply commercial fertilizers. Drilling these with the spring seeding, as is common practice, provides soluble salts for plant nutrition at the same time when the seed is equipped with its own stored organic supply to take off on its mission of growth without any fertilizing help from the soil. By mid-summer, when the plant is growing rapidly and its roots are down in the deeper, more moist horizons, the fertilizer is left far behind in the dried surface soil where the crop's roots are not active. Or those fertilizer salts may have reacted with the soil which has immobilized much of them. Even a good share of them may have been taken up

Neglecting the maintenance of the organic matter in the soil under continuous corn (upper right photo) for half a century means (a) no natural winter cover crop, (b) a soil surface dashed into slush and erosion by a single rain (lower right photo), and (c) little infiltration and reserve moisture to let this surface soil be hotter by ten degrees during the summer, in constrast to the soil also in continuous corn but given 6 tons of manure per acre annually. (These plots on Sanborn Field were photographed on Friday in late April—upper photos: plowed on Saturday, rained on during Sunday, then photographed again on Monday—lower photos.)

by the microbes. Such share is larger according as the soil is richer in the more stable forms of organic matter. The microbes are the first crop we grow every spring. They are the first to feed on the inorganic fertility applied in the presence of soil organic matter. They are also the first crop to be shocked, as we might readily expect, by drilling a heavy dosage of a highly concentrated salt into a limited soil volume.

Nature uses no concentrated salts in building up her soils to an ecological climax. Instead, she uses the decaying organic matter of the past dead generations of exactly the same plant species which she is growing. In the forests, nature applies her organic matter on the top of the soil. Tree roots are perennial. They do not die annually to leave a dead generation of themselves as fertilizer for the next generation. Since they grow on less fertile or more highly developed soils they make mainly wood. This is a low value in terms of organic fertilizer and a soil treatment which is only slowly active. Forests have, therefore, built up but little soil organic matter and there is only a shallow surface soil layer. With the trees shading the forest floor, the organic

Deeper prairie soils of only moderate degrees of development under lower rainfalls in the midcontinent had a big reserve of organic matter as their "constitution" for survival in spite of, rather than because of, our maintenance of their organic matter, as this Barnes loam illustrates.

By putting fertility down deeper into our soils, we invite roots down to put more organic matter there as these roots of sweet clover testify. It is not the deep-rooting crops that make deep, open, fertile soils, but rather the reverse.

matter accumulated on it contains considerable moisture during much of the season. But since its woody nature makes it a poor microbial diet, it cannot decay rapidly to be much of a "fertility turnover." But yet its nutrient delivery rate in cycle increases as the season gets warmer. The trees' growth processes are consequently fed a bit faster thereby at the time when the mounting temperatures increase their life processes and their need for fertility. During the later season with its lowering temperatures, the reverse of all these processes into lowered rates for the winter takes place. This occurs because the ceasing decay of the organic matter shuts down the delivery of fertility from the organic sources. The inorganic or mineral-decomposing processes close down or drop to a very low rate too. Forest trees exhibit their growing season because of the decaying season of their own litter on the forest floor.

In the case of the soils of the prairie, nature applies her organic fertilizer mainly within the soil. Lightning and consequent prairie fires may burn the organic fertilizer she might return to the top of the soil. Prairie grasses and their roots are mainly annuals. They die back each year to leave much of themselves as organic matter to be added to previous accumulations, supplementing the larger supply of reverse rock and mineral matter of more active potential in the soil. Grown on such soil of only moderate degree of development with the prominence of legumes in the natural flora, these organic fertilizers are more rapidly decayed. Therefore, the seasonal decay is delivering rapidly not only carbonic acid from the woody part, but also nitrogen, phosphorus, potassium and the other many inorganic essentials in the grass which served so well to grow the beef in the buffalo. The decay is also feeding the microbes with organic compounds to grow their own antibiotics, hormones, etc., by which they protect themselves and contribute organic compounds to the plants, also by which they may be healthier themselves. It was on such soils where the buffalo was healthy too without any ministrations from man in his professional classifications.

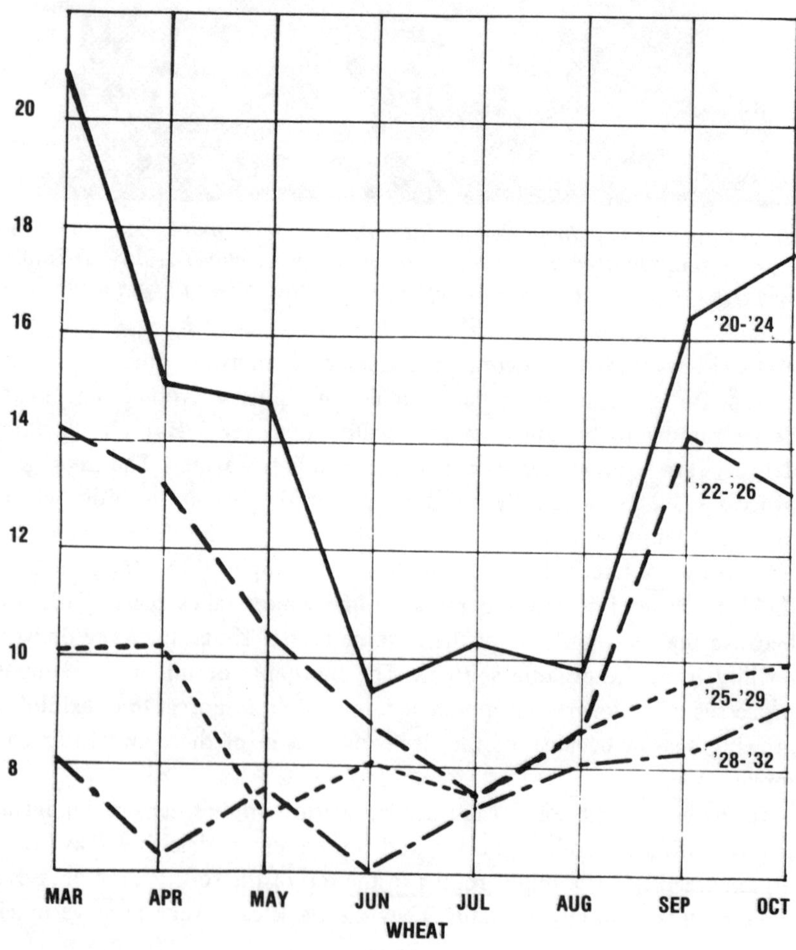

Measurements, at regular intervals, of the nitrite nitrogen supplies built up from the decay of the organic matter within the soil during the growing season, in such virgin prairie soils put under cultivation, illustrate how well the seasonal rise and fall of the rates of decay and delivery of active inorganic and organic fertility is turned and synchronized to meet the increasing and decreasing seasonal needs for fertility by the crop growing above the soil. The crop growth tells us that it results from the effects of the temperature on the soil for this decay service delivering nutrition. We have commonly been crediting the crop growth to the direct effects of the temperature on the

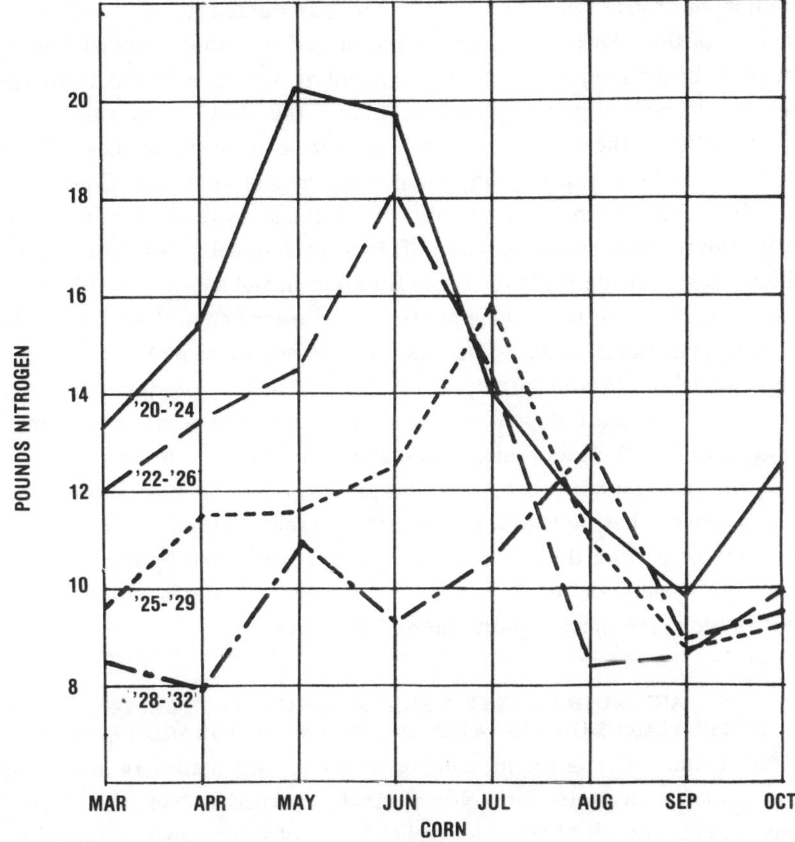

The accumulating rise of the nitrate nitrogen content of the soil, and its decline through the crop's removal of it during the growing season, give one form of curve for wheat (see graph on previous page) and a different one for corn (graph above). In either case, the curves are at decidedly lower levels; and for corn, the maxima drop markedly with continuous cultivation as depletion of the soil organic matter continues.

Crops Must Grow Their Own Soil Organic Matter   107

plants. The removal of the nitrates by the crop according to its needs for them for growth—which is in a different part of the season in the case of wheat, for example, than it is for corn—give us evidence for this fact. The decline from the heights to which nitrates accumulate as the season's maximum with the succession of the years while the soil is under continuous cultivation also testify accordingly. We are slow to realize what the organic matter within the soil is doing when we concern ourselves so fully with the yields of the crops and so little with the whys thereof.

Unfortunately, we have not paid much attention to, much less have we understood, the services to our crops by the natural organic fertility in the soil when we plowed out our virgin sods. At the outset, they were not so productive nor the crops so healthy. But after what amounted to a few years of compositing of those surface residues in the turned-over sods, the yields of healthy crops began to mount. These were aided by correction of the fertility deficiencies through application of lime and all the other inorganic fertilizers. The benefits of these resulted in no small measure because they were supplements to the excessively organic diet of the soil microbes. Consequently they fanned the microbial fires to burn out the soil's reserves of the organic matter all the more rapidly. The shift to a mechanical agriculture and more tillage increased the draft for those fires within the soil. Now with the economic pressures on the land mounting, and their demand for larger yields per acre to demand such also per man, we are not apt to make close observations of the health and quality of all the agricultural produce. We do not realize that these are dependent on a high fertility in terms of the organic parts as well as the inorganic, or so-called mineral part of the soil.

The crops of plants as construction workers cannot operate wholly on mineral supplies. They depend on the wrecking crews passing on much secondhand material from the soil organic matter grown there by previous crops. The more secondhand material duplicating much of what is required for new construction, the more rapidly the building goes up.

ANTIBIOTICS ARE PROTECTIVE ORGANIC COMPOUNDS
OF THE PLANT'S OWN CREATION ACCORDING TO THE SOIL FERTILITY

Plants and animals in the wild must grow, and doubtless always have grown, their own protection against attacks by viruses, bacteria, fungi, insects, worms and all of these lower life forms of infections and parasites of them. In nourishing themselves to make more crop of body weight, they seem to grow resistance, immunity, or whatever we call the plant's and animal's means of protecting themselves against other smaller forms of life that would survive by slowly consuming them. That the chemical compounds which the plant and animal create as protection are proteins, or much like it, is suggested by our use of blood serums as inoculations to grow our own immunities as protection. Such protectors in the case of some fungi have now been isolated as substances which we call antibiotics. Since animals and man must eat proteins to make most of their own body proteins for growth and protection, the converse of that fact suggests that the attacks by infections, parasites and pests may be due to deficiencies of the proper proteins in the nutrition of the animals, and those due to the deficiencies in the fertility of the soil by which the plants grow the proteins required for their own protection.

One might also be led to infer then, that when the plants are growing a bigger crop, they are also growing more protection against all these troubles with diseases and pests. Such may or may not be the case. Increased crops of vegetation may be due to merely a higher ratio of carbohydrates to the proteins formed in the plant because of the unbalanced fertility treatment for it. Consequently, the loss of proteins as protective ones may result from this imbalance in the plant's own created products because of imbalances via the soil. Excessive nitrogen applied as an imbalance may also fail in giving pro-

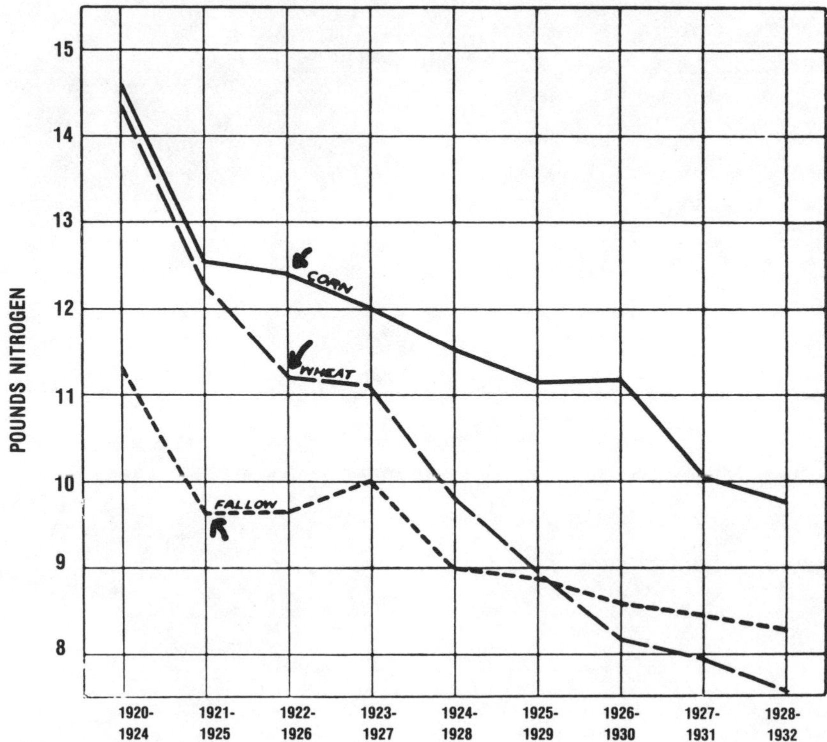

The averaged levels of nitrate nitrogen in the soil during the growing season go lower and lower with time to tell us that the organic fertility of the soil is being burned out.

tective proteins against viruses, bacteria, etc., even when it makes the crop grow taller and greener and contain more crude protein. The nitrogen may be taken into the plant from the soil without being converted into the specific proteins made up of the amino acids required as suitably protective ones. Measurements of the amino acids composing the proteins in timothy hay showed that fertilizer nitrogen applied alone gave more yield per acre. But it gave less concentrations of the many amino acids, and especially of those known to be essential for the nutrition of man.

Plants build up bulk readily by synthesizing the carbohydrates. The increase in bulk depends also on the synthesis of some protein. Increased yield of bulk can be obtained by merely diluting the protein with the carbohydrate. We have failed to recognize this as what happens when the soil fertility for

growing protective proteins declines but holds up for growing the crop yield as carbohydrates and fattening value, but low protein and lower protective value. We are not able to tell which amino acids make which proteins to be protection against what kind of plant and animal troubles, or specifically, against insects, fungi, bacteria or viruses.

Now that we have been using nitrogen fertilizers more generously and have highly demonstrative stimulation of rich green growths of plants, we credit

Nitrogen at a low level of fertility in the soil (10 M.E. per plant) left this spinach crop open to insect attack, though less so as there was more calcium applied to the soil (left, lower to upper photo). But nitrogen at the higher rate (20 M.E. per plant) helped the plants build their own protection at any calcium level (right, lower to upper photo).

much to the nitrogen. We forget that we have been building up the soil fertility generously in many other plant nutrients while aiming to help legumes grow nitrogenous fertilizers. Consequently, when we are fertilizing with only nitrogen, we are apt to believe that we are growing more crude protein. We might believe, therefore, that we are also growing more protection against disease and insects. But that the qualities of the proteins are not necessarily protective ones is suggested by the demonstration of borer attacks on ears of corn grown with different soil treatments and kept under observation for over two years.

The use of nitrogen only as fertilizer treatment of the soil, which showed no major deficiency save nitrogen by soil tests, grew only small ears of hybrid corn in the very dry season. Under observation in storage, this ear was gradually eaten by the lesser grain borer. Laying against this one so highly infested by this insect, there was the larger hybrid ear representing the yield from the plot fertilized with both nitrogen and phosphorus. Borer attack on this larger ear from the larger grain yield was limited to the table. The open-pollinated ear of corn kept under observation during only the last four months in contact with the infested one had but one or two holes indicating borer damage.

These observations of the two hybrid ears of corn suggest that a bigger yield was also better protection against this grain pest. It was bigger protection not because more nitrogen (by which we characterize protein) in the soil grew more crude protein, but apparently because a supplement to the nitrogen, namely phosphorus, balanced the plant's functions to use the nitrogen to make the specific protective proteins. Here it was the difference in the soil fertility as difference in plant nutrition that meant the survival of the species via protection through probably some organic compound in the seed.

BETTER ANIMAL HEALTH DEPENDS ON ORGANIC FERTILITY TOO

We do not know all that needs to be known about building soils for fullest nutritional service to crops, but there is encouragement from such observations to try feeding them not only to grow big crop yields per acre, but also big protection, hopefully, against virus, bacteria, fungi, insects, worms and other pests destroying our crops and livestock. This protection is connected with organic compounds in the microbes, in the plants, in the animals and in man. It is connected, also, with the organic matter in the soil nourishing the microbes and plants as the initial synthesizers of protein from the fertility elements and organic compounds in the soil. There is much yet to be done via soil management for animal health that will put the basic science of agriculture under the statement from the art of agriculture, which has long ago told us that "To be well-fed is to be healthy."

CHAPTER 10

The Trace Elements, More Soil and Health Relations

THE SO-CALLED TRACE ELEMENTS have come into particular concern as inorganic essentials for the health of microbes, plants, animals and humans during the last decade. They are now under research studies at most every agricultural experiment station. They are also offered for sale not only in fertilizers but also as feed and food supplements extensively in various combinations on the lists of proprietary medicines, even though they are more nearly foods than drugs. Their deficiencies as well as their essentiality in many life processes are being recognized more and more widely.

Trace elements are required in amounts so small that they are commonly considered only a trace in chemical language. Their essentiality could not be established until some particular chemical methods and the more delicate biochemical ones of sufficient sensitivity were developed. If their presence is essential, their magnitudes required in mere traces are not of so much concern. Their significance lies in the fact that their absence means death, or the failure to be healthy and to reproduce normally.

THE LIST OF THEM IS ENLARGING ITSELF

As essentials for the growth of crop plants, manganese, boron, zinc, copper, chlorine and molybdenum make up the list commonly included in the trace elements at this moment. Cobalt is essential for the microscopic, blue-green algae. It is also essential for animals. Chlorine has long been known to be essential for animals. It was added to the trace elements essential for plants only recently. Iron has long been known to be required in trace amounts for plants, hence if classified on this basis, it should be included in the list to make eight trace elements so far as present knowledge has established them for growing plants and microbes. Whether boron and molybdenum are needed by animals and humans is not yet determined. Iodine is required by animals and man but not necessarily for plants. Outside of these last three, the other trace elements are known as essential for both animals and man.

For many years only ten elements were listed as requisites for plant growth. Three additional ones made up the list of thirteen required by animals. Dr. Cyril Hopkins, one of the pioneers among soil-plant scientists of the first decades of this century, grouped the symbols of the ten for plants by the letters of his own name as follows: C-carbon, H-hydrogen, O-oxygen, P-phosphorus, K-potassium, N-nitrogen, S-sulfur. Then by omitting the I-iodine for modesty's sake and adding Ca-calcium, Fe-iron and Mg-magnesium, the combination read "C. Hopkins Cafe, Mighty good," as a convenient memory help for retaining the list in mind. As essentials for animals, the I-iodine was inserted and Na-sodium and Cl-chlorine were added. With carbon coming from the air and hydrogen and oxygen from water and nitrogen in the ultimate also from the air, though commonly from decaying organic matter in the soil for non-leguminous plants, the rest of the elements are contributed from the soil for the processes of creation of all life.

The list of trace elements is still an increasing one. Chemical analyses of plants and animals always reveal a long list of inorganic elements present in the ash. We must keep our minds open to add other elements to the present list when all life forms must be considered. Fluorine is regarded an essential element by its presence in the enamel of the teeth, but yet we do not know the physiological functions in which it serves. Silicon is found in the hair, the nails, animal hoofs and body portions duplicating the hair, for example. Vanadium has been suggested as a replacer for molybdenum in plant growth and is having attention for many possible functions since, like molybdenum as a soil treatment, both are demonstrated in crop differences when applied as ounces per acre. Bromine has been suggested as substitute for chlorine. This growing list of trace elements is telling us over and over that the inventory taking, by chemical analyses, of the ash of plants and animals does not necessarily reveal all nor exclude any of the elements and their essentiality, either as delicate parts of the living tissue or as tools in the life processes growing them, protecting them or reproducing them.

SINGLE-CELLED LIFE MAY BE EITHER POISONED OR FED BY TRACE ELEMENTS

As the simplest forms of life, the microbes are very sensitive to trace elements either as deficiencies or as excesses. In fact the poisons we have used to destroy microbes in many instances, have demonstrated their essentiality in trace amounts in the growths of these very microorganisms. Variation in the microbial growth reflects differences in the concentration of trace elements in the test medium more delicately than such is shown through measurements by chemical methods. As a consequence, there have been developed the microbial methods of testing for the presence and amounts of the trace elements. These biochemical methods have laid the foundation for our research on them. It is the fungi which have been used as testing means for a long time. It was these lowly (even despised) forms of life which opened the research in, and production of, antibiotics. It is the simpler, black-fruiting mold called *Aspergillus niger* which serves in many soil tests of the concentrations of several of the trace elements there. It is a biological test using life itself as the measure and with a delicacy that transcends other methods.

Molds live by their digestion or decomposition (we call it rotting) of other organic matter both dead and living. But that fact should remind us that they use organic matter by direct absorption and synthesis of it into the organic compounds of which their cells are composed. Their speedy growth can be readily expected then, when most of what they use as material of their construction of themselves is already highly fabricated toward the compounds eventually composing the final mold tissue itself. Mushrooms are illustrations of one of the most rapidly growing fungi. They build themselves as a special organic matter of food value for us through but little simplification required from the decay of the organic matter in the manure compost through which the mushroom's mycelial, or thread-like growth literally runs. The organic compounds of certain stages of the compost's decay are merely catalyzed, as it were, into another set of such compounds we consume in the growing mushroom. This growth of the mycelium takes place in the composted manure of the spawning bed before the "cashing soil" covers it as mineral source for the fruiting or spore production of the mushroom.

It is the trace elements which are usually chelated. That is, they are connected into organic combinations for their services in chemical activities. Their main role is that of catalysts. They are the tools serving to give speed to chemical reactions. They may be combined with some organic substance and by that union some additional chemical reactions may be initiated. The fungi are very good illustrations of such and their services in the very simple forms of life. They serve to reduce our thinking to the single cell scale. Even in our own bodies of their myriads of cells, we need to be reminded that we are living and healthy because each cell performs its many functions, its separate part of what makes up the functions of the body as a whole. Each cell must then have its small share of the essential elements, both major and trace, in quantities too small to be measured, but yet large enough to be the difference between death and life biochemically.

The molds reflect trace element deficiencies readily by a reduced rate of production of their mycelial mat or their fibrous growth. They reflect these deficiencies most pronouncedly in the failing reproduction, that is, in their crop or yield of spores. These single celled, or even the few celled forms of life illustrate what is true for plants, animals and man, namely the reproductive process is the first one to show by its irregularities or failure that the general nutrition supporting that delicately adjusted integration of many complex biochemical reactions is not fully undergirded by a truly complete nutrition at all times. Interruption, possibly only an irregularity, in the reproduction processes, whether in plants, animals or man, is one of the first symptoms of nutritional deficiencies going back to the soil in most instances.

That fact may well prompt many cases for research with not only the trace elements but with fertility in its larger scope. Soil deficiencies in the trace elements may well be the cause for the invasion by the bacteria accompanying the failing reproduction in the cow. Erroneously, then, we might believe the bacterial presence the cause of the abortion, the weak calves or the dwarfs. We might be prone to set up a national campaign to fight the bacteria when in reality they are merely another symptom or a consequence of the nutritional deficiency as is the failing calf crop. These two, *i.e.*, the bacteria and failing calf crop, are contemporaneously but not necessarily causally connected. Fighting the bacteria is a common cause in the faulty nutrition. Fighting the bacteria is apt to be just a fight without a complete diagnosis of the reasons for the presence of the bacteria. Trace element deficiencies or other shortages in a complete nutrition, and consequently the invasion by microbes as an early step in the impending disposition of a prospec-

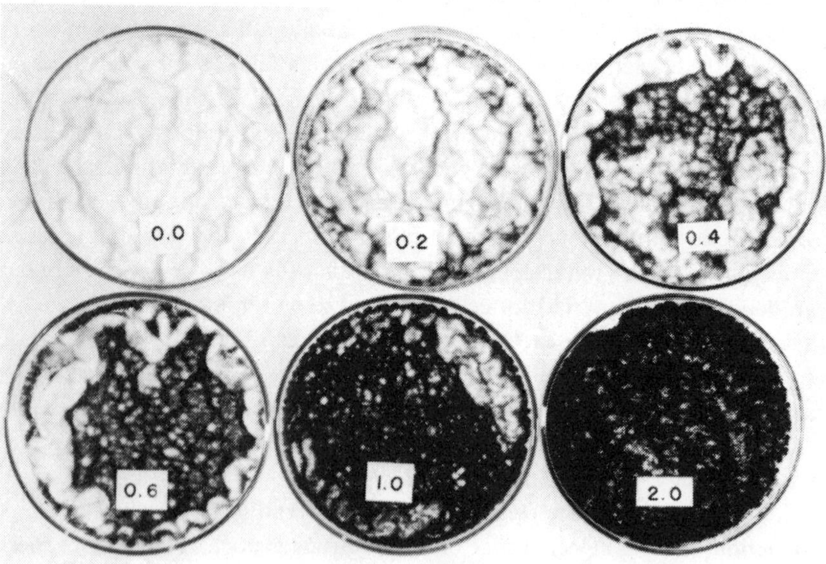

Nitrogen at a low level of fertility in the soil (10 M.E. per plant). Increased spore production or more fecund reproduction (darker color) by the fungus Aspergillus niger *reflects the increasing copper, viz. 0.0 to 2.0 parts per million used as the test scale for measuring the copper in soils.*

tive cadaver, is a bit of thinking apparently too morbid to make us more proficient in keeping animals as healthy when we feed them as they are in the wild or when they feed themselves.

OUR OBSERVATIONS ARE OFTEN CORRECT, BUT OUR IDEAS AS TO CAUSES AND CURES ARE MORE OFTEN INCORRECT

Biochemical behaviors under our own ministrations may often be disturbed because of the error in our interpretation or explanation of them. The effects by copper used as a fungicide in sprays of Bordeaux mixture have long been explained as due to a poison, to copper which kills the fungus by contact. It seems high time to doubt the validity of that explanation. It seems to be more in accordance with the facts and behaviors to believe that the soluble copper sulfate, converted into one of the most insoluble forms of copper as an hydroxide on reaction with the soluble calcium oxide, could not be much of a poison by penetrating the cells of the fungus. It would seem more logical to believe (a) that the colloidal or gluey copper hydroxide sticks to the leaf, (b) that the carbonic acid of the leaf surface slowly converts it into the more soluble copper bicarbonate, (c) that this compound is then absorbed into the leaf tissue and (d) that thereby it makes up the nutritional deficiency of this element which may be required for the plant's synthesis of its antibiotics by which it protects itself against a fungus invasion. This seems a logical postulate when it was the fungi demonstrating their antibiotic protection against other fungi through which the idea of antibiotics was discovered. This may be merely a theory, but such also is the idea of copper hydroxide as a contact poison; and a very poor theory in the face of the most insoluble copper hydroxide.

Copper sulfate as a vermifuge for stomach worms in sheep rests on a similarly doubtful explanation of its functioning. Copper treatment for worms causes them to loosen their attachments to the stomach walls and to pass out of the alimentary tract as living—not dead—worms. They are apparently not killed by the copper acting as a poison for them even when used in the very soluble sulfate form. Again it would seem that the presence of the worms might be a symptom of nutritional deficiency of the trace element copper in the sheep's diet. Drenching the sheep with copper and correcting the deficiency may let this animal build its antibiotic protection and prohibit the worm's attachment to the mucous walls by which these pests are excreted. We need to ask, "How do wild sheep know their medicine to stay healthy?"

That copper is required to make black sheep grow black wool in place of changing it to a gray color was established some years ago in Australia and New Zealand. That fact should suggest copper deficiencies in the soils of Missouri when it is the common report that "Sheep born black soon turn to a

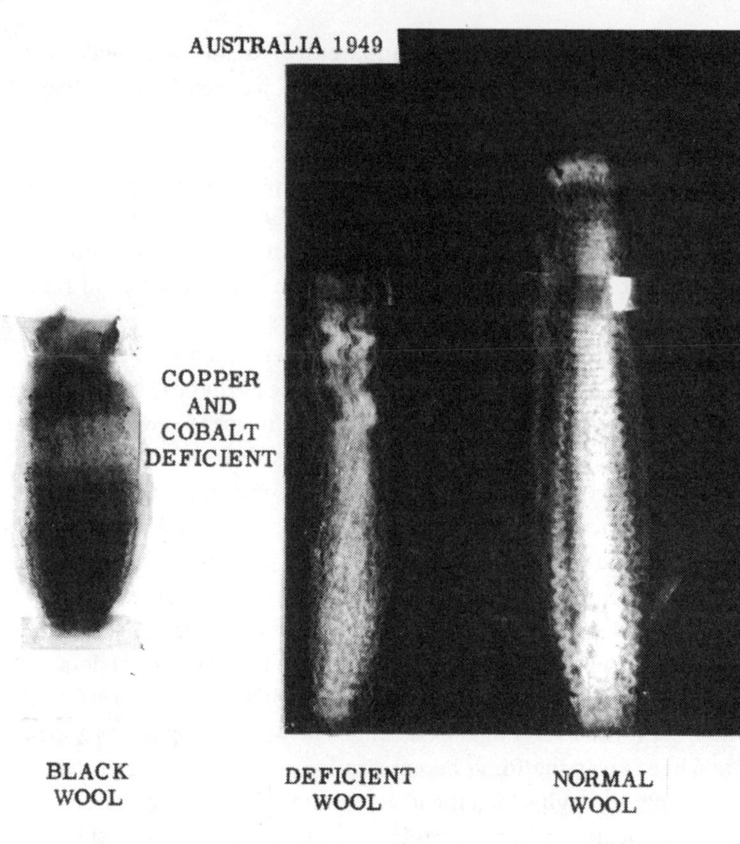

The white wool specimen with long waves at its top, in place of short ones (center), was taken from a sheep suffering copper deficiency and then drenched with it to change the abnormal to normal wool as shown in such a specimen (right). Black wool (left) shows color changes as the sheep was shifted about on different grazing areas.

gray white.'' We need to start giving a few black lambs copper drenches occasionally in lots in contrast with other black lambs given no copper. That could be a simple biological test right on the farms in various parts of the state to learn whether the soils growing the feed for those lambs commonly changing from black to gray are deficient in the trace element copper. Such a test is far more delicate than any chemical one yet applied to the samples of the soil. By combining our thinking with the animal's biochemical behaviors, we may fit both the animals and ourselves into nature's potential for our mutually better survival through better nutrition via more fertile soils. Trace elements may be both poisons and protection when we know more about proper use of them in terms of their functions in the body processes.

118   Soil Fertility and Animal Health

Such functions by copper will not be measured and correctly interpreted by ash analyses of amounts present in the younger, and then again, in the older life forms. The quantity of copper is not a readily measured variable when a small amount of this element removes the symptom of gray-white wool of the sheep by growing the wool out in black color and then prohibiting the return of the symptom of copper deficiency in gray-white wool again for several months, even if copper is withheld. The introduced copper is retained and reused by the body. It protects the body for a long time. It is not a daily requirement in the feed. It is not a variable except when withheld over a long time in the life of the animal. This is an influence quite different than we commonly expect from nutrient elements.

It is most logical, then, to view the trace elements not as constituent parts of the mass of tissue grown, but rather as tools used repeatedly in the processes of life. They are probably entering into a biochemical reaction to bring it about even more speedily, and then coming out to repeat the reaction on another mass; just as we envision the helps by catalysts modifying immense masses but themselves never consumed or retained in them.

One would scarcely expect trace elements to demonstrate much effect, then, in the plant's production of carbohydrates or in the animal's laying on of fat and other equally gross and common manifestations of what are biochemically less complicated processes. Rather the proteins, so generally manipulated by many enzymes, would suggest themselves as the realm of physiological processes within which the influences by essential elements in trace amounts might be recognized. Enzymes are combinations of protein-like molecules, vitamin-like ones, and inorganic ions (very often the trace elements). If enzymes by the hundreds are responsible for creating the proteins, composed as they are of their score or more of amino acids, then it would seem well to postulate that the variation in the supplies of trace elements coming from the soil might cause variations in the proteins or in the array and amounts of the amino acids synthesized from the elements only by the plants and microbes.

If one hammer and one saw are essential tools for constructing a dog kennel, the conclusion does not follow logically that there will be a correlated increase in the number of saws and hammers needed or serving in constructing a hundred kennels or a dog house a hundred times as large. Nor would an inventory of these kennels and dog house find a single hammer or a single saw necessarily within them at the close of the building performance. Trace elements must be viewed as tools or as enzymes which do not conform to common correlation-thinking where large masses are concerned and are necessarily connected with large cause controlling their behaviors. In classifying the trace elements for their functions, the criterion of increased quantity is not fitting. There is a call for a new criterion, namely, they must be classified according to their role in the biosynthesis of the proteins from the carbohydrates or in the higher quality of the crop as nutrition in terms of proteins and all else associated with them.

TRACE ELEMENTS AS CATALIZERS OF THE PROTEINS SHOULD DEMONSTRATE THEIR EFFECTS IN THE LEGUMES MORE THAN IN THE NON-LEGUMES

If, therefore, the trace elements serve more extensively where many enzymes are active, we should expect soil treatments which are adding trace elements to be less demonstrative through the non-legumes than through the legumes. This should be anticipated, especially if we make analyses for the proteins, inorganic elements, vitamins and all else associated with the proteins which we need to produce more extensively. Some preliminary studies were undertaken in which trace elements were used as fertilizers for timothy given the basic soil treatments but then divided into plots treated with nitrogen in combinations with the separate trace elements. Chemical analyses of the timothy for the inorganic elements including the traces indicated concentrations of these in good amounts as one might expect even for some legumes. When the timothy was assayed for the ten essential amino acids, it was lower in concentrations of seven of those than is common even for red clover.

Had one compared the ash analyses only, one could not have believed the two hays, *i.e.*, the timothy treated with trace elements and the clover given various treatments, so very widely different. The differences in the amino acid contents suggested the significance of the clovers as a better feed because of their higher concentrations of the more commonly deficient amino acids. That the clovers were a better feed was clearly demonstrated when the timothy was under bioassay by feeding it to weaning rabbits in 1954 and the heat wave came along to kill the rabbits with about equal fatalities for all of the timothy hays, regardless of trace element and nitrogen treatments. These fatalities amounted to 70% of the original numbers. The experiment was repeated under the extension of the hot summer until 30% of the rabbits were again killed off. Then the timothy hay was replaced by the clover hay only to have no more fatalities, regardless of the high temperatures. Chemical analyses of the ash told little and the assay of the amino acids suggested more about the differences between the timothy fertilized with trace elements and the clover as feed for rabbits. But only the feeding of the hays to the rabbits in testing their nutritional services could tell us that the hay from the non-legumes, even if fertilized most fully, including the trace elements, would not keep the rabbits alive under high temperatures while the red clover would. Fertilizing timothy with the full soil treatments including nitrogen

Inorganic Elements in Timothy

Treatment (Lbs./A)	P %	Ca %	K %	Mg %	Na %	Mn p.p.m.	Fe p.p.m.	B* p.p.m.	Zn* p.p.m.	Co* p.p.m.	S p.p.m.
No nitrogen	.175	.280	1.43	.210	.07	147	355				1460
40# nitrogen**	.123	.217	1.18	.210	.045	147	364				1250
40# N + 30# N**	.129	.204	1.24	.268	.055	134	295				1270
40# N + 60# N**	.129	.187	1.16	.222	.045	119	419				1230
Mean	**.139**	**.222**	**1.25**	**.227**	**.054**	**137**	**358**				**1302**
40# N + 60# N**	.129	.187	1.16	.222	.045	119	419	3.7	10.4	.06	1230
40# N + 60** + B	.128	.235	1.11	.275	.055	134	345	5.5	4.8	.08	1280
40# N + 60# N** + Zn	.120	.192	.97	.262	.060	60	295	5.2	20.8	.02	1500
40# N + 60# N** + Mn	.136	.193	1.08	.157	.050	105	537	5.9	14.4	.08	1270
40# N + 60# N** + Co	.130	.221	1.03	.275	.050	105	352	6.2	12.4	.03	1410
40# N + 60# N** + Cu	.130	.238	1.08	.281	.050	105	337	6.0	22.4	.03	1420
40# N + 60# N** + all 5 trace elements	.152	.240	1.07	.238	.045	60	375	5.9	32.8	.03	1300
Mean	**.132**	**.215**	**1.07**	**.250**	**.051**	**98**	**380**	**5.5**	**16.9**	**.05**	**1344**

*Determined by spectrographic methods.
**Nitrogen fertilizer was applied in the form of solution.

Ten Essential Amino Acids and Nitrogen in Timothy
(mgms/gm dry matter)

Treatment (Lbs./A)	Nitrogen	Methionine	Tryptophane	Lysine	Threonine	Valina	Leucine	Isoleucine	Histidine	Agrinine	Phenylalanine	Total Amino
No nitrogen	9.15	.50	2.48	.715	2.60	2.86	12.1	6.35	.812	2.86	2.48	33.76
40# nitrogen	8.84	.39	2.16	.894	2.00	2.64	9.0	5.11	.652	2.68	2.02	27.55
40# N + 30# N*	10.2	.35	2.24	1.20	2.40	2.76	11.5	5.58	.595	2.88	2.09	31.60
40# N + 60# N*	13.4	.52	2.90	1.65	2.80	3.69	14.6	7.12	.917	3.75	2.95	40.90
Mean	**10.4**	**.44**	**2.44**	**1.11**	**2.45**	**2.99**	**11.8**	**6.04**	**.744**	**3.04**	**2.38**	
40# N + 60# N*	13.4	.52	2.90	1.65	2.80	3.69	14.6	7.12	.917	3.75	2.95	40.90
40# N + 60# N* + B	14.7	.70	2.90	1.82	3.26	3.74	14.6	7.05	.801	4.00	2.80	41.67
40# N + 60# N* + Zn	13.0	.56	2.86	1.58	3.04	3.23	13.2	7.13	.824	3.45	2.66	38.53
40# N + 60# N* + Mn	11.5	.57	3.39	1.40	2.80	3.48	13.5	7.20	.675	3.88	2.81	39.70
40# N + 60# N* + Co	13.4	.76	3.39	1.66	3.04	3.56	14.6	7.42	.893	3.88	2.86	42.06
40# N + 60# N* + Cu	12.1	.45	2.73	1.68	2.78	3.12	14.2	6.62	.670	2.55	2.53	37.33
40# N + 60# N* + all 5 trace elements	13.5	.72	3.61	1.84	3.06	3.59	13.6	7.31	.892	4.00	2.94	41.56
Mean	**13.1**	**.61**	**3.11**	**1.66**	**2.97**	**3.49**	**14.0**	**7.12**	**.810**	**3.64**	**2.79**	

*Nitrogen fertilizer was applied in the form of solution.

Nine Essential Amino Acids and Nitrogen in Red Clover
(mgm/gm dry matter)

Plot No.		Essential Amino Acids									Total
	Nitrogen	Methionine	Lysine	Threonine	Valine	Leucine	Isoleucine	Histidine	Agrinine	Phenylalanine	Amino Acids
3	15.72	1.04	4.03	3.53	4.21	4.40	3.45	.97	9.38	3.28	34.29
25	16.72	1.44	5.15	4.64	5.26	6.08	3.91	1.12	8.57	4.37	40.54
26	14.70	1.08	4.65	4.07	4.02	5.40	3.94	1.35	7.77	5.88	38.16
27	17.91	1.15	5.35	4.79	6.37	7.50	4.74	1.40	10.30	4.75	46.35
28	18.53	1.22	5.80	4.39	5.71	5.75	4.14	1.37	8.87	5.10	42.35
34	17.55	1.33	4.38	4.23	5.59	5.45	3.95	1.15	8.80	3.87	38.75
35	18.18	1.33	5.40	4.61	6.54	6.40	4.74	1.42	9.90	3.97	44.31
36#	19.01	1.55	6.50	4.93	6.12	7.00	5.03	1.70	11.30	5.04	49.17
37	15.44	1.48	5.60	4.92	6.59	6.75	5.61	1.50	11.30	5.38	49.13
38	16.70	1.50	5.40	4.77	5.79	5.74	4.16	1.40	9.07	5.36	43.19
39#	20.91	1.25	5.65	4.73	6.54	6.28	5.05	1.66	10.20	6.51	47.87
Mean	**17.4**	**1.31**	**5.26**	**4.51**	**5.70**	**6.07**	**4.43**	**1.37**	**9.59**	**4.86**	

 # Plots Nos. 36 and 39 were given trace elements as soil treatments and also had the crop residues returned regularly.

served to dilute the concentrations of most of the inorganic essential elements by increasing the plants' vegetative output. But with more nitrogen—and this in combination with the five trace elements—much was shown for an increased concentration of the inorganic elements in the hays.

These facts illustrate again the principle established in some research by Dr. H. E. Hamptom at the Missouri Station, namely that according as the plants are higher in protein will they take the inorganic elements off the clay colloid more completely than the same plant species will at lower levels of its protein production. If the element, deficient for complete protein production by the plant, is supplied to let the plant suddenly make its more complete protein, the plant root so nourished represents a greater energy gradient

Soybean plants showing varied growth with all nutrients constant except boron. This was supplied at the rates of 0, 13.1, 26.2 and 52.5 parts per million (left to right). The leaf specimens are from a comparable position on the plants (left to right).

from the fertility elements held on the clay of the soil to the root's contents into which they move more extensively. It is the greater movement of fertility into the more proteinaceous crops which helps them to be richer in protein. By the same token, the legumes and other high protein crops exhaust the soil's inorganic fertility supplies faster and to a lower final level in the soil. Trace elements, as the neglected ones, suggest that they are a more common deficiency than we yet realize in the larger problem of growing enough crude proteins to say nothing of making those more complete in terms of the required amino acids.

Because trace elements suggest their role in connection with the plant's biosynthesis of proteins, researchers have used the soybean plant more re-

Soybean plants showing varied growth with all nutrients constant except manganese. This was supplied at the rates of 0, 3.6, 7.3 and 14.7 parts per million (left to right). The leaf specimens are from a comparable position on the plants (left to right). Their symptoms are suggestive.

cently. The seeds of this legume represent complex if not complete proteins. The seeds include much beside the many essential amino acids required for both animals and man. As yet, we have not been very successful soil managers to get bigger yields of the high protein soybean seeds per acre by fertilizing the soil as we have been in getting bigger yields of the high carbohydrate, low protein corn grain per acre for agriculture in general. In the latter, *i.e.*, the production of the carbohydrate by the corn, the plants' photosynthesis represents their use of the sunshine with an efficiency of about 30%. In the former, *i.e.*, the production of the protein in the seeds, the plants' biosynthesis represents the same use with an efficiency of scarcely 3%. Thus we might expect non-legumes to pile up the carbohydrates about

Soybean plants showing varied growth with all nutrients constant except magnesium. This was supplied at the rates of 400, 800 and 1600 parts per million (left to right). The leaf specimens are from a comparable position on the plants (left to right). They demonstrate the symptoms of the deficiency of magnesium.

ten times as fast as the legumes pile up the proteins, or in terms of bulk produced, "legumes are hard to grow." Protein production is the result of the living tissue creating it and not one of merely collecting it from the air and water under synthesis by sunshine power. Trace elements in the process using only 2% of the sun's energy in making the proteins cannot be as demonstrative, via bulk delivered, as can magnesium in chlorophyll using 30% of the sunshine power for piling up the carbohydrates in fuel and fattening feeds.

Trace elements used in experimental studies of the growth of soybean plants to test the significance of boron and manganese, in contrast to that of magnesium, a major element, when all else of known requirements was supplied, demonstrated clearly that the plants use them effectively from such very small supplies that they may well have been neglected before we had means of measuring their significance in plant and animal nutrition more refinedly. The health of the leaves is particularly significant, showing itself as different sizes and of deeper green colors with the increase of the trace elements in only parts per million.

Alfalfa, another legume on which we are depending more and more to cover our problems of the proteins for our animals, is demonstrating that deficiencies of the trace elements are disturbing the functions of this crop more commonly. Alfalfa is being pushed beyond its natural ecological areas or out of the climatic settings which develop the soil to be more nearly balanced nutrition for alfalfa. It is grown more on the sandier soils where the low supply of exchangeable fertility, because of low clay content, does not give all that this protein-rich feed needs to make it such. Some trials in Texas using ordinary superphosphate and that supplemented with a trace element mixture demonstrated the significance of the latter so far as chemical analyses suggested, but more so when the cattle exhibited their preference if given the opportunity to demonstrate. The data illustrates the increased mobilization of the many fertility elements into the crop beyond those applied as soil treatment. When we fertilize the soil we mercly offer the fertility at higher levels in the soil. That is no guarantee of their higher concentration within the crop, save in general where the elements applied have been moved from a limiting factor to one more nearly in balance for the plant's processes.

PLANT SYMPTOMS SUGGEST THE SOIL'S
SHORTAGES OF TRACE ELEMENTS TOO

Just what particular symptoms a plant will show when a certain trace element is deficient has not been so specifically worked out. Such has not yet been done even for all the major elements. But, nevertheless, some suggestive symptoms may well be tabulated. It would be well to remind ourselves that the symptoms of trace element excesses or toxicities are even more lacking.

Iron shortage is one of those recognized earliest in history in the chlorotic leaf pattern and the yellow to white development of the intervenal tissues of the leaf. The vascular tissue may remain green for only awhile. Intervenal striping of cereals may precede the complete bleaching.

A deficiency of copper, like that of iron, affects young growths. In fruit trees it is called *summer dieback* of the shoots. Chlorosis, wilting, cupping of the leaves and breakdown of the leaf margins occur and then spread to the midrib.

Shortage of zinc has recently been widely recognized, especially in newly irrigated districts of the Pacific Northwest where its diagnosis found it to be a

| | Fertilized With | |
	Superphosphate Only	Superphosphate and Trace Elements
Protein Total %	23.8	24.4
Protein Digestible %	16.8	17.3
Calcium %	0.580	0.527
Phosphorus %	0.189	0.286
Potassium %	4.74	5.56
Magnesium %	0.211	0.27
Sodium %	0.175	0.207
Sulfur %	0.114	0.136
Copper p.p.m.	5.1	11.2
Cobalt p.p.m.	Trace	1.0
Manganese p.p.m.	0.2	1.1
Iron p.p.m.	225.	210.
Zinc p.p.m.	7.0	17.0
Molybdenum p.p.m.	Trace	0.4
Boron p.p.m.	7.7	13.3
Iodine p.p.m.	151.0	221.0
Total Ash %	11.85	12.95

Trace Element Mixture Applied at 200 Pounds Per Acre

Borax	200 lbs.
Copper Sulfate	80 lbs.
Iron Sulfate	300 lbs.
Manganese Sulfate	400 lbs.
Cobalt Sulfate	20 lbs.
Bonemeal	1000 lbs.

baffling plant hunger. The leaves of reduced size, often chlorotic, for beans, corn, citrus, apples, pears, etc., are spoken of as *little leaf disease* when it is, in reality, a starvation. It is a disease due to malnutrition, as many so-called diseases truly are. The symptom was most commonly associated with alkaline or neutral soils and was confused with a nitrogen hunger or protein shortage. Zinc sulfate has become an important fertilizer not only in the northwest but also in the southeast.

The visual symptoms from manganese deficiency vary from necrosis by blotches on the leaves from chlorosis to breakdown of the seed cotyledons, especially of legumes. The cabbage and beets show molting as chlorosis in the older leaves. In oats the gray speak is the symptom. In barley the small brown spots along the leaf blades are common.

When molybdenum shortage came to light, it was shown in cauliflower's failing to develop the leaf parts to correspond with the size expected from the lone midrib. It was called "whip-tail." Distorted young growths, death of growing tips with abnormal elongation of older leaves are common. In legumes the deficiency shows up in the failure of the nodule bacteria to use atmospheric nitrogen even with nodules present.

Boron deficiency has had extensive notice and prevalence. The meristematic tissues are affected so that a kind of "bushy type" of alfalfa plant results. Some crops have *heartrot,* or there is *brown heart*, or *cracked stem*, or *corky core* according as the crop is sugar beet, swede and turnip, celery or apples respectively. Many others have been listed and the deficiency magnified by the drier seasons.

Chlorine deficiency was only recently shown for the tomato plant. The wilting of the tips of the leaflets with chlorosis later is a common symptom. The failure of the tomato fruit is a severe symptom.

All of these symptoms emphasize themselves at the points of most active cell processes or growths. Naturally they are often connected with the proteins by which alone the growth is brought about. Unfortunately the symptoms may be so evident so late that little of remedy is possible. Yet in many cases, the mere spraying of the plant allows enough of the essential nutrient to be absorbed by the leaves and get into quick function for correction of the trouble. All of this raises the question whether the trace elements, like copper used as poisons to kill lower plant forms like fungi, are not doing so via their absorption into the plant for its better nutrition and own antibiotic protection rather than any poisoning effect on the invading organism. Use of the trace elements suggests that we are merely ministering to the health of the plant and are removing or preventing the disease symptoms by the more complete nutrition of the plant. By that support, the plant protects itself under our management precisely as it must do if it survives with health in the wild.

TREATING SOILS WITH TRACE ELEMENTS MAY IMPROVE ANIMAL HEALTH ALSO

A growing number of baffling animal ailments exhibit themselves not as a single symptom but in a list of them. These usually duplicate a part of the list in each one of a series of cases. Such facts suggest that the breakdown is, in some general processes, brought on possibly by malnutrition. Deficiencies of proteins active in so many processes (growth, protection and reproduction), would (a) not to be expected to exhibit the same set of symptoms, (b) not be localized in the same body part and (c) not be limited to the same function in every case. The protein-starvation disease among humans in Africa, known as kwashiorkor, illustrates those facts well when its complicated symptoms are not yet necessarily all tabulated, but when a speedy cure is effected by feeding the fat-free, dry-milk proteins.

Inquiries at a recent regional veterinary meeting elicited many reports that ailments of the animal's skin and excessive irregular secretions of the mucous membrane were probably the most commonly baffling now in certain areas. If we remind ourselves that the mucous membrane as lining inside, and the skin as covering outside the body are both of the same embryological origin, we are also reminded that they are the most extensive secretory and excretory organs of the body. When these functions break down, this occurrence is evidence that the internal functions of which excretions and secretions are only manifestations, have already broken down or the very processes of life are failing. Should we expect a single, localized symptom when the visible functions of these protective areas, as are the mucous which must handle all the ingested compounds and the skin which makes all external contacts, may be failing?

Additional replies emphasized the breakdown of the functions of the liver and of increasing portions of its tissues at earlier age. We have been slow to appreciate this excretory and secretory organ which is the body's chemical censor of digested substances before they are allowed to enter the bloodstream. Nature does not throw anything into that circulating medium so directly as we do in using the hypodermic needle to bypass the liver's censorship. That the liver has its troubles even with our feeding of animals by normal methods, was demonstrated by the marked differences in the appearances of the livers of the different rabbits (little mates) sacrificed for such observations after experimental feedings on corn grown with different fertilizer treatments with and without the trace elements.

Complaints about this organ as a food portion of the carcasses of both smaller and larger slaughtered animals are increasing. As a chemical guardian, the liver must not only remove useless and detrimental compounds coming from the digestive tract, but it must also synthesize many for which the starter substances can come only through nutrition. As a storer and user

The soil fertility growing the grain is reflected in the differences in (a) the color (black, then yellow, encircled upper row) of the livers from rabbits fed on corn grown on soil with no treatment and (b) the texture when fed on it grown with fertilizer (broken liver, soft texture, encircled, middle row). They were firm and uniform in color when grown with fertilizers, including trace elements (Missouri Experiment Station).

of copper, one of the trace elements, the liver plays a major role. Other trace elements and their possible management by the liver have not yet been catalogued. Malnutrition and deficiencies may be causing breakdown in secretory and excretory services—our major protectors against damage by irregularities in nutrition—by our body's organs in which the proteins are the motivating actors. They are the physiological protection. For this in its multitudinous phases connected with the hundreds of enzymes, the trace elements as coactivators and coenzymes are playing major parts not yet fully comprehended.

Not only in the failing protection of the body but also in the reproduction as propagation of the plant and animal species should the importance of the trace elements be appreciated and more intensely studied. Some research by the Missouri Experiment Station in 1946-1949 with dairy herds which were failing to conceive and to calve, used extensive soil treatments with both major and trace elements on 330 acres to demonstrate the restoration of the health and production by means of better animal nutrition "from the ground up." The changes in the animal behaviors by this procedure alone were most encouraging through the basic approach by starting with the process as a natural one originating in the soil.

Before the soils had been treated, the herd had only six conceptions from 47 matings of 24 cows. The first year after the soil treatments were applied, there were 37 conceptions and 85 matings of 49 cows. Then for the second crop year in sequence to the extensive soil improvements, there were 63 conceptions by 63 cows from 167 matings. During the three years, the percentages of cows that conceived were 25, 75 and 100 consecutively, as the result of treating the soil—the starting point in creation.

At the outset of these studies the lowest semimonthly milk sale per cow was 151 pounds, and that in the first half of January. Three years later, the corresponding figure was 306 pounds coming in the first half of October.

The low level of vitality of the newly born calves exhibited itself before the soils were treated, in the 12 abortions and the eleven calves too weak to survive, with 41 cows calving, or only 18 calves from this number. Two years after the soil treatments, there were 58 normal, strong calves and seven abortions from 65 cows calving.

At the beginning of the study, 29% of 79 head (23) of breeding age were reported as infected by *Brucella abortus* (Bang's disease). Three years later (with the herd kept in contact) only 20% of the 98 breeding cattle were reported as positive reactors. By the end of four years, 17 heifers from the herd had calves and both mothers and offspring were always negative to the test and supposedly free from infections by *Brucella abortus*.

These facts tell us that our explanations of the causes of (a) the carrier and (b) the associated microbes should be recognized as connected with the failing, or the deficient, nutrition when we consider the microbes as the causes of the disease. We use this clinical evidence even to destroy the animals themselves in an environment of our own creation, and one quite different from that in which the animal would survive in the wild. We accept as causes what may be results of the more deeply-seated causes with which we are not familiar. The carrier and the associated microbe in epidemics have been fittingly characterized as "The clinical bull in the ecological china shop." The microbial invasion may be a kind of scapegoat or a case of calling in the pathology to explain—and to destroy the evidence by slaughter—where the perverted physiology and the faulty nutrition are unknown.

A DECLINING SOIL FERTILITY REFINED TO TRACE ELEMENTS PORTENDS MORE, NOT LESS, TROUBLES IN ANIMAL HEALTH

Trace elements mean more problems of nutrition. Our knowledge of soil fertility has made signal progress. We have changed our appreciation of liming the soil as a fight on soil acidity to one of stocking the soil with the major fertility element, namely with calcium, for growth with proteins concerned. When that element needs to be present as an exchangeable and ionically active one in amounts of 6,000 pounds per plowed soil layer, this is a figure gross enough to be measured. It started our appreciation of what made up the list of major elements. Essential nutrient elements have been tabulated on the list and measured in amounts as large as that figure and as small as one-sixteenth of an ounce per acre. Both calcium and molybdenum are required to grow the legume clover successfully with three tons of the former element (7.5 tons limestone) and one-sixteenth of an ounce of the latter serving to illustrate the variations as quantity by which essential elements control quality in terms of being able to grow, to protect and to reproduce the food crops we need.

These facts about soil as nutrition, resulting from studies on the delicate biochemical manifestations of plants showing deficiencies, are bringing us to be concerned about deficiencies as the cause of what is so readily accepted as diseases and dismissed from careful physiological research. When all the essentials from soil as nutrition for life are not yet known qualitatively, we cannot be too emphatic in discussing the known ones quantitatively. Quantities in the ash are a matter quite different than are the quantities in action of all the essentials brought together in the soil and moving under their own squad behaviors into the root hair and the process we call plant growth.

We must be ready to add more elements to the list of essentials. Their quantities are not of significance when fractions of an ounce per acre spell life or death in one and tons are required in the case of another. That situation challenges our ability to diagnose with accuracy in many cases. It suggests that in place of using postmortems, we might more effectively turn to preventions in the form of complete nutrition as the animal in the wild outlines it, rather than as we outline it for castrated males imprisoned in a feedlot. The increasing numbers of degenerations of body functions, still classified as diseases with the implication that by drugs we can "cure" them, ought to convince us that as we mine the fertility of the soil more, we are merely perverting the life stream more nearly to its own extinction. If it has been flowing in spite of us and not because of us, we must go back soon and build up the soil as the starting point of Nature's Assembly Line of Creation by attention to its fertility, lest we might have no original standard left on which to premise our thinking of just what a soil must contain if its fertility is to give us truly healthy animals, as when they themselves do the growing, the protecting against diseases and the reproduction with fecundity sufficient for safety against extinction. The trace elements are a challenging part of the soil fertility, when by such small amounts they may be the authority deciding what lives and what starves.

MYSTERIES OF CREATION

The mysteries of creation haven't all been put under button pushing technology. And the trace elements are a part of it. I had a letter from a man in Florida. He had read something about my remark that we aren't including an inventory of all the elements that are nutritional. He wrote, "Albrecht, I'm growing citrus down here. I graduated from Purdue." He said, "Tell me what to put on the soil as trace elements. I'd like to try it." I gave him a gunshot. Copper, manganese, zinc—I gave him a list. Come Christmas, he sent me some fruit. Earlier he wrote. He said, "We're having trouble keeping our dairy cows out of our citrus grove." He said, "Every morning that herd of cows goes through the fence. Now we have a strong fence," he said, "but we have one cow that goes right through it anyway. She's in every morning." So I said, "Don't laugh at that critter. She's just a little smarter. She's an A student." Anyway, I received this citrus. I put the fruit in my wine cellar. When we finished that grapefruit bushel, the last two in the bottom had been cracked. But they hadn't spilled any juice. They had split so that the slices separated out. And I checked the shipping date and it had taken six weeks for them to come to my place. This was interesting because the fruit hadn't been taken by the green mold, and citrus turns green in a hurry if it isn't properly fertilized. I think the copper he had put on that land at about 5 pounds per acre protected against the green mold. Later he wrote that he had corresponded with some 100 experts. Not one thought we knew what we were doing. "But," he said, "I just closed my contract for all of my citrus at a nice markup because I had a better fruit."—*Dr. William A. Albrecht, in an interview with Acres U.S.A.*

CHAPTER 11

The Problems
of the Proteins

THE PROPER FEEDING of livestock to keep the animals healthy is mainly a problem of providing the extra essential proteins as supplements to the homegrown carbohydrate-rich crops. These energy providers usually give good yield on humid soils but insufficient proteins and other essentials coming along in them. Those climatic areas of soils developed under higher rainfalls, and under fertility depletion by that natural part of the climate, have always had their problems of protein deficiency in their forages. Those areas have been the main reason for the commercial feed business. They gave us the whole problem of purchased protein supplements. Consequently, the economics of getting those supplements is the major factor hindering the health of livestock.

In hoping to make cheap gains or cheap gallons, we are not depending very much on managing our own soils to grow or to create the products bringing those gains and gallons for us. But rather, we are shifting to the hope of buying the health via the market's offerings of feed. We cannot, thereby, make our efforts dovetail into those put forth by the animals in their choices of feed and in their struggles for their own better health when they are grazing on mixed herbages lush growth. That is what nature offers in the wild during the spring season in which the births of the myriads of life forms are naturally scheduled. The purchases of supplements extend the protein lifelines from the soils which are deficient in growing this essential and all else, including both major and minor essential inorganic elements and special organic compounds of even medicinal services. Buying supplements lowers our view of the creation of domestic forms of animal life to the level of a simple economic transaction of costs and market prices; when for healthy animals it must be one of physiology, nutrition, health, reproduction and the creation of life by beginning with that potential in the soils which we cultivate supposedly for that high purpose.

We are apt to forget that agriculture is first a natural biological performance, and second a financial transaction in the marketplace with the death
of those biotic products rather than with their health and the survival of their
respective species in prospect. If we have lost sight of the life of agriculture in
our management of it, should we not anticipate considerable bad health and
disaster when we are concerned with only its economics and the technologies
applied to it? Bringing the livestock from birth to the slaughterhouse is a
biological matter proceeding according to the laws of Mother Nature. It challenges our more careful study. Once the animal is slaughtered and life
eliminated, then it becomes an object of technology and economics which can
be fitted readily into the demands of these two man-made controls and manipulations as we might choose them.

DEFINITION OF PROTEINS—A PROBLEM OF CONFUSION

Just what all is included in the term *proteins* is not a clearly comprehended
matter by even the sciences of organic chemistry and biochemistry. Proteins
have always been characterized by the fact that they contain nitrogen combined with carbon, hydrogen and oxygen. For that reason the nitrogen in
many organic compounds has served to label them crudely as proteins. Unfortunately, this element nitrogen may take several chemical forms within the
organic substances we grow as crops. As a truly protein form it occurs as one
atom of itself combined with two of hydrogen; and this so-called amino
nitrogen or protein-characterizing unit is combined with a particular carbon
atom in an organic chain compound of more carbon, hydrogen and oxygen
alone. The measuring of this kind of nitrogen necessitates burning the carbon and hydrogen off to catch the nitrogen in the form of ammonia or in its
combination with three, rather than two, atoms of hydrogen.

Other combinations of nitrogen which are not chemically combined so specifically, and which do not serve in the processes of nutrition for growth, protection and reproduction of life forms as the amino nitrogen does, are also
converted into ammonia by ignition in this chemical determination. Since the
different compounds supplying the amino nitrogen vary widely in their percentage of nitrogen and since we have been content to say that proteins contain an average of 16% of this element, then any forms other than amino
nitrogen bring much error into the determination of the proteins. We are
labeling as *protein* all the nitrogen regardless of whether it might be ammonia, nitrate, nitrite and other forms which the growing plant takes from
the soil and we trust will be converted into amino nitrogen and function truly
as a life-carrying protein. In feed analyses we have been selling organic and
even inorganic nitrogen, but were believing we have thereby offered protein
supplements.

The legume forages, and parts of the seeds such as milling remains or seed cakes from which the oils have been extracted, were the pioneer concentrates as protein supplements to the hays and bulky carbohydrate matters of feeds. But these are no longer offered so extensively on the open market for economic transport to great distances now that livestock production has become a more ambitious part of agriculture in most every farmed area. Proteins are a problem right at home on every farm. Milling products don't move very far from the mill. Those grown in Kansas formerly moved as wheat to the eastern mills. They were the protein supplements for the New York City millshed. They served in feeding operations near that metropolis for its meat supply. Now those milled products move as commercial feed in bags. Soybeans are also moving but short distances as protein supplements for livestock feed. They are moving away from animal into human consumption where the problem of protein supplements is rapidly shifting from one of animal products to vegetable forms. We are seeing the problem of even crude proteins confining its solution to the soil areas where the soil fertility grows the proteins more commonly. The acuteness of the problem is emphasizing the soil as the responsible factor under the crops.

Biochemistry is removing some of this confusion as to what proteins really are. Research has shown that the body's needs for all that is included in the term *proteins* consist of the many demands for the separate components of proteins, namely for the amino acids. These are the units into which the proteins of foods are separated for their higher solubility, greater simplicity and effective absorption by the processes of digestion. We are clearing away the confusion connected with the crude proteins measured by the amount of nitrogen multiplied by 6.25. We are now viewing the problem of the proteins as one of providing the specific amounts of each amino acid required to keep the body in balance so far as intake and outgo of nitrogen are concerned. These include ten for the white rat and eight for the human. Then also, the provision of many of the others in the two dozen now known is required for their contributions to body processes not yet tabulated, even for only nitrogen balance. It is this knowledge built up by study of body physiology that is outlining this problem, not as one of only the crude protein supplements, but one of providing the essential amino acids. It is becoming necessary to know the kinds and the amounts of them in the feeds, and thereby the quality of the homegrown and readily available protein, in order to supplement it with those amino acids in which the array of those on hand is deficient. We are no longer concerned with the nutritional balance as simple as one of carbohydrates versus crude proteins. Instead we are concerned with nutrition balanced with respect to the required amino acids which those crude proteins provide in relation to the carbohydrates.

The confusion resulting from our satisfaction with the crude protein contents of crops in certain ratios to the carbohydrates, as we make up our animal feeds, has allowed the declining soil fertility to bring more and more crops into use which may still have considerable nitrogen, but much of that may be in non-protein forms. Even much of what is protein may be deficient in two or three of the ten (or eight) essential amino acids. Mining our soils for larger yields of bushels and tons has consequently resulted in lower and lower quality of the forage and feed proteins with respect to their shortages in tryptophane, in methionine and in lysine. These are the three amino acids which we usually hope we get when we purchase protein supplements.

We have not been testing the feed crops on the farm for their contents of tryptophane, for example, which has a unique chemical structure. In its carbon chain part there is one amino nitrogen unit, or the truly protein form of this element. The human body digests this chain section out of the amino acid and uses it when we eat tryptophane. Another part of this essential amino acid is a carbon ring form. This ring contains another nitrogen atom but not an amino form in it. Unfortunately this entire ring remains undisturbed by digestion. The ring nitrogen is not used by the body. Instead, that nitrogen is just so much deception by 100% when we measure its total in tryptophane and consider it all as protein or amino nitrogen. This much nitrogen of even an essential amino acid passes unchanged through the digestion to be excreted as part of either the compound indole or of the slightly more complex skatole—both giving the common fecal odor. When we measure crude proteins by ash analyses for the nitrogen as the index of proteins, then, we are not measuring even the true proteins correctly. Much less are we taking an inventory of the essential amino acids in the crops we grow. We add much confusion about what feed quality we are growing in our crops when we know no more than that about the proteins in the supplements we purchase, and then combine the crops and the supplements and believe we have solved the protein problem. By the increasing incidence of disease among our cattle, resulting from that confusion, those poor beasts are paying for our ignorance of the nutritional values which our soils and crops are creating in terms of the essential amino acids in the proteins.

Methionine as the second commonly deficient amino acid among those essential serves to supply sulfur in the protein form. It is related to cysteine, also a sulfur-carrying protein but not necessarily an essential amino acid. Unfortunately sulfur, which is no more prevalent and no less deficient in our soils, in general, than phosphorus, has had little attention as a fertility element. It has been included unwittingly in commercial fertilizers when sulfuric acid is applied to phosphatic rocks to make their phosphorus more soluble and active. Sulfur-containing crops like rape and all those allied to the cabbage, and like garlic and others in the onion group, were once considered important feeds and animal medicines. The latter are taken by preference in animal choice; even the garlic in the spring time to suggest the animal's verification of the medicinal values claimed for these pungent plants in connection with some baffling animal diseases, including brucellosis. This commonly missing amino acid should cause our concern about sulfur in the soil's fertility store, particularly when experiments applying more sulfur have resulted in more methionine in some crops to suggest that these can be grown more nearly complete in protein by proper treatments to build soils accordingly.

Lysine, another commonly deficient amino acid, is not distinguished by any chemical structure widely different from most of the other amino acids, save that it has two amino nitrogen units. One is in the customary structural position while the other is at the end of the six carbon chain opposite the acid end. Like the tryptophane and the methionine, the lysine is deficient in the cereal grains. Fortunately all three are not so low in soybean meal or sesame meal, both of which represent possible concentrates for animal feed. They are all three much higher in certain fish, hence fish meal may be more significant for health than we realize for both humans and animals. Since fish must feed, in the last analysis, on the vegetative crop of the sea, and since they may need to eat the organic components which they assemble into their body proteins much the same as higher life forms must, there may be much value in putting some of the particular seaweeds into the animal diet. When the cattle in pastures along the sea coast wander to the beaches to search out and eat some of the vegetative inwash, they may be demonstrating their refined choice there as they do of herbages with different fertility treatments. They may be finding differences in quality of the proteins as well as differences in the amounts of iodine and other trace elements in the kelps and other marine vegetation to which we would give emphasis by this unique search.

Unfortunately in choosing the crops we grow, we have paid no attention to the quality of their proteins in such a great detail. We have appreciated legumes for their higher crude protein concentrations. But when they will not even grow on many soils, unless those are carefully fertilized, even the legumes are making up less and less of our homegrown feeds. We are not holding up the protein supply from that source. We are increasing the shortages of commonly deficient amino acids in the feeds. We have believed that legumes grown on most any soil were taking nitrogen from the air, when perhaps they were growing like a non-legume on the soil nitrogen or were even losing seed-nitrogen back to the soil. With our choices from a variety of crops merely to grow them, we may be magnifying the confusion relative to the quality of feed protein we are producing for animal health and animal reproduction.

PROTEINS PRESENT A PROBLEM
OF BALANCED SOIL FERTILITY FOR PARTICULAR CROPS

The words *soil fertility* represent more than you might believe from what is commonly reported as the contents of the fertilizer bag. That term includes all that comes from the soil to contribute in any way in the growth of plants. This divides itself into (a) the inorganic and (b) the organic parts. The inorganic divides itself further into (a) the major and (b) the minor or trace elements. Then each of these groups may be divided again into (a) the cations, or positive electric charges by which they are active, and move toward the cathode or negative pole of a battery circuit and (b) the anions, or negative electric charges by which they are active and move toward the anode or positive pole. In the services by the inorganic cations in plant nutrition, we visualize them weathered out of the rocks and minerals as soluble, active ions; then they are *adsorbed* on the clay which results simultaneously from the weathered rock; and while not taken from there by water, they can be exchanged from there by any other positively charged ions. For that, the hydrogen resulting from the carbon dioxide excreted into the moisture around the root serves as an exchanger. This non-nutrient then takes the place on the clay for the nutrients made active to move them into the root for plant nourishment. Thus we get a clear vision of the activities of cationic fertility elements like calcium, magnesium, potassium, ammonium, iron, copper, zinc, manganese and sodium in accordance with chemical laws. According to these, they may be held in the soil against loss from there in pure water percolating through the soil. But yet, they may be taken by the plant root using sunshine power to give the carbohydrates respired by that root into the carbonic acid surrounding it and activating the soil fertility into plant nutrition.

Unfortunately, however, while we envision the clay of the soil as the large sluggish, negative colloidal molecule holding and exchanging the above list of nutrient elements of positive charge, we do not comprehend too clearly just how the nutrient anions, like nitrate, chloride, iodine, sulfur, boron, phosphorus and molybdenum are held and exchanged into the plant root. Plausible theories concerning them have been formulated. Improved management of plant nutrition may result when they are fully established as the facts.

The organic fertility is another division of plant nutrition; the services of which research has not yet elucidated clearly. Chemical science has been less specific in its interpretation of organic reactions than it has in its elucidation of those inorganic. As a consequence, much controversy has arisen regarding organic matter in soil and in fertilizers for its services in plant nutrition in contrast to sevices from chemical salts. The organic compounds taken up from soils by crop roots for plant nutrition are the seriously neglected other half of soil fertility. Their production and conversion by soil microbes for increased plant growth with the advance of the season may mean more in the final composition of the proteins in the crop than we appreciate.

When mushrooms, living almost wholly on decaying organic matter, are the most rapidly growing food crop we produce, should we not believe that crops might be more speedily grown if there were more organic matter in the soil serving in their nutrition? Would it not be possible to postulate that compounds as complex as the amino acids would be more quickly synthesized by crops if those organic creators could take up from the soil some organic compounds, like indole and proprionic acid for example, the constituent parts of tryptophane, which is also called indoleproprionic acid, and thus synthesize this essential amino acid more speedily than if only the elements carbon, hydrogen, oxygen and nitrogen are the starting materials offered by the soil, water and air?

The grazing cow taking grass, microbes, and soil all into her paunch as a highly active and fermenting mass suggests the important role microbes play in converting organic matter from a chemically stable compound like cellulose, into digestible food values for either the plant root or the cow's alimentary canal. The cow's paunch suggests her literally direct connection, thereby, with what resembles the soil and the organic matter of its production under microbial transformation at body temperature. It suggests the preparation for later digestion along the alimentary tract for introduction into the liquid protein stream, which her blood represents in nutrient balance with the entire body protein. Should it be a dangerous stretch of the imagination to visualize the cow's blood as the combined organic and inorganic chemical expression by that warm-blooded body, of what a protein potential for animal life the soil fertility growing her grazing really is? We have been slow to believe that the critical chemical testing of the levels of organic and inorganic essentials in the cow's blood might give helpful suggestions on soil treatments, ministering more directly to good animal health via preventions of diseases than drugs and potions do when they are aimed at the cure of health deficiencies. That soil treatments for growing more nutritious feeds can be surmised from analyses of the blood for improvement of animal health and for even prevention of diseases has already been demonstrated both abroad and in the United States.

Proteins as the body's protective agents against disease can be grown and chosen by, or administered to, the grazing animal as its medicine in the form of nutritious feed only when the fertility of the soil is balanced for the particular crops by which those protein requirements result from their biosynthetic activities. Such proteins are problems of the proper choice of the crops and the proper choice of the fertility treatments on the soil. They are problems of all this good judgment coupled with the maximum help the animal can give by its behaviors to our critical study of even its bloodstream or other body properties as they give suggestions about good health via complete nutrition for it.

For too long a time have we had a fear of microbes. We have shunned the changed organic products resulting from the many microbial digestive activities. We forget that the cow's paunch is an outstanding case of her use of the many changed organic products the microbes can bring about within that voluminous organ. As a result, she, quite different from the non-ruminating pig and chicken following her and requiring so-called animal proteins in their rations, can live wholly by vegetarian principles. She does not require animal proteins.

We forget that microbial activities also in our foods contribute to their preservation and even improvement in nutritional values in terms of vitamins and other special compounds when we use fermentations for preparation and preservation of sauerkraut, wines, beer, sauerturnips, sour milk, cheese, and other foods with distinct flavors. Microbes in the soil have even come into recognition for the antibiotics they make. We have not yet appreciated the microbial flora and fauna of the soil for their services in transforming crop residues, plowed under, into not only organic fragments serving directly as organic nutrition of the crop but also into the chelating agents for moving the inorganic fertility into the crop for more effective plant growth. All of these services emphasize the microbial connection with the plant's production of the proteins. Nitrogen connects itself with the protein production of plants by coming even from the gaseous supply of it through the help of legume bacteria. But that occurs only when all inorganic elements of fertility are completely assembled and in proper ratios. Proteins are a problem in our health program because even organic fertility must be balanced too if we are to grow this food component which has not yet become a product which the industrial assembly line can put out.

**BIOSYNTHESIS OF PROTEINS BY CROPS
VIA SOIL'S ORGANIC REMNANTS OF DIGESTED PROTEINS**

The synthesis of the more commonly deficient amino acids, like tryptophane, or of the plant hormones like indole-acetic acid, for example, may require that the plant take up from the soil some organic molecules of larger dimensions excreted by animals from their digestion of amino acids. This was suggested by observations of the varied growth behaviors of Michigan dwarf bean plants. Dr. Francis M. Pottenger of Monrovia, California grew these ordinary navy bean plants in some abandoned cat pens as a second phase in his experimental studies of the differing nutritional values of (a) condensed, (b) evaporated, (c) pasteurized and (d) raw milks for cats during two years. These were combined with a constant allotment of cooked food as the diets for all of the cats.

The pens had been filled with an infertile, well-washed, nearly-pure quartz sand. This was fertilized by only the urine and cat dung buried by them during the two years of these critical studies of cat nutrition. The pens were arranged according to the one variable in the test, namely, the kind of milk fed the cats. They were separated, however, according to male and female sex of them.

After the nutritional phase of the study of the cats was completed, the pens were left abandoned for a time. They were observed, then, the tremendous differences in the volunteer weed crops of a single species according to the different treatments of the milk represented in the feed for cats. All the urine and dung from the condensed, evaporated and pasteurized milks apparently did not put into the sand enough fertility even to invite weed growths, save for the pasteurized milk through the males. The raw milk of the same composition as the original of the pasteurized, had put so much back even after feeding the cats better in terms of physical vigor and persistent reproduction, that the weed growth filled the pens completely.

In order to observe these matters more closely, the weeds were removed and the pens planted each with two rows of beans. The first significant fact demonstrated was the different growth behaviors of the bean plants—even if all represented the same lot of seed. Wherever the cats' diets contained the heated milks, the plant growth characters were those which one would call the bush or dwarf kind of bean. Where the buried dung from the cats fed the raw milk was the fertilizer, the plants were like pole beans with vines climbing the screened sides as high as six feet. Here were decided differences in the growth behaviors of the plants, not because of the claims of the plant breeder, but contrary to them and because of heat treatment of the feed (in some not to exceed 140 F. for half hour) going into the animals making the manure which fertilized the soil growing them.

Still more significant was the observation of the fecal odor in the bean seeds harvested from the bush type of plant growth in the pens of cats fed on heated milks. There was no such odor detectable from the seeds harvested where the buried dung was from cats consuming the raw milk or where the plant growth was of the pole bean type.

Since indole and skatole are the common urinary and fecal excretions

A

B

Dung buried in the sand by cats reflects its differing manurial values under the volunteer weed crops according as the male cats were fed (A) evaporated milk or (B) unpasteurized (raw) milk. The manurial values under dwarf beans from (C) evaporated milk were less startling than those from (D) unpasteurized (raw milk) which changed the plant growth characters from "dwarf" to "pole" bean, regardless of the same pedigree of all the seed.

resulting from digestive changes of tryptophane, one of the essential amino acids, it would seem mere hallucination to believe these powerfully odoriferous compounds to have resulted from the plants' regular creation of them from customary soil fertility elements. It is more logical to believe that some of the excretory organic compounds from the cat's digestive or from the microbial transformations were taken up by the plants—after burial in the soil—and were transmitted into the seed without enough analytic or synthetic change to remove the odor. With the odor so characteristic to suggest no breakdown of the ring combination of the indole or skatole, there is much reason to believe that this is the correct route of travel of the organic compounds of protein potential undergoing little change as these self-announcers reported to the nose at the outset and the finish of their apparent journey from the dung in the soil to the seed or the bean's reproductive part.

The chemical stability of indole and skatole, because they are ring compounds of carbon including nitrogen, is more reason to believe in their movement within the soil into the plant and into its seed without chemical change. Few if any common microbes can break down the indole and skatole ring. Microbes or digestion may break the carbon side chain off from the ring when the combination of these is tryptophane. One needs only to be reminded that indole-acetic acid, that is the indole ring with an acetic acid side chain, is the common plant hormone giving pronounced growth of roots and of shoots extended into vines. This change in type of plant characters would not occur unless this larger molecular structure is absorbed into the plant cell. Indole taken up from the soil may become this hormone, that is the indole-acetic acid, through but little chemical change represented by the addition of a two-carbon chain to the indole ring. With not much more change through the addition of amino-proprionic acid—a three-carbon chain with the amino nitrogen—it becomes tryptophane.

If this is the plant's performance of starting its tryptophane resynthesis by using the indole going into the cat dung from the digestion of tryptophane, it suggests that indole might be the starter organic compound also for the plant's synthesis of the indole into the plant hormone changing the "bush" bean to a "pole" bean. The presence of the indole odor in the bean seeds, and its absence in others, suggests indole uptake for the plant's possible conversion of this remnant of tryptophane back into it as a part of the bean species' survival through complete protein in the seed; provided a little heat treatment for pasteurization—without any removals or additions—does not destroy some of the accompanying organic or inorganic matters. These may well be necessary accompaniments of the indole for the biosynthesis of that again into the tryptophane by the plants. If these syntheses are prevented, then only the indole apparently is put into the bean seed as a fragment and not as the complete protein. Tryptophane may be more fundamental in the minutia of reproduction, like the genes and chromosomes, than we yet believe.

BIOTIC VALUES IN ANIMAL EXCRETA TRANSCEND THOSE CHEMICAL

From such observations of the many integrated phenomena in the soil, in the plants and in the animals, there comes the suggestion that one of our most commonly deficient and essential amino acids may depend on specific organic compounds in the soil fertility for the synthesis of it as animal feed by our crops. There is the further suggestion that the required organic compounds result from the digestion, excretion and return to the soil of such original amino acid by animals or microbes. That return keeps parts of it, like organic ring compounds, in a cycle of organic matter going from the soil into protein via the plant's synthetic activities and then from the protein in plants and animals back to the soil via animal digestion and excretion or in microbial wastes. There is the implication that we have not appreciated that cycle as possibly one of even very specific organic compounds when, in the absence of man, nature builds a single crop to a glorious climax of pure stand without diseases and insects prohibiting. It implies further that the problem of the proteins may be a sin of our own omission of animal manures from the list of fertilizers for creating the proteins by which so much is done to guarantee animal health through good nutrition from the ground up. Perhaps the problem of the proteins is also a sin of commission when we accept the concept of proteins by measuring only the ash nitrogen and apply this element in salt fertilizers for crops with our contentment and belief in their creating crude proteins thereby. Unfortunately the sickening animals must render the verdict regarding our moral code in connection with the management of our soils and our crops in their behalf for keeping them healthy.

A VISION OF PLANT NUTRITION

For about 25 years I've worked on this Epsom salts business. Frequently, after a hernia is repaired, bowels won't move past that hernia. So they give Epsom salts. And if they check, they'll see that urine is throwing the protein out of the blood. Protein is wasted because the Epsom salts ruin the membrane in kidneys and keep them from doing their normal work. When you take Epsom salts, that salt replaces the calcium in the wall of your intestines and it throws everything it can because that membrane is no longer normal. It just throws everything from the bloodstream till it flushes it and can go back to your bones to get some calcium to rebuild intestine walls. When I gave that to Dr. F. M. Pottenger, he said, "You've got a good theory because if we've got a highly rheumatic person and give him Epsom salts, he's so low in calcium he throws the calcium out so badly that it kills him." Now the medical profession knows that they shouldn't give Epsom salts, but they do. But you see with this hernia, the kidney wasn't functioning when the magnesium went through. The magnesium that the bloodstream had to throw out through the kidney was knocking the kidney. Now here's my theory. Now remember my work. If I didn't have my soil loaded high enough with calcium, the nutrients were going from the plant back to the soil exactly the way they go from an intestine. If I don't have this calcium-saturated soil high enough, the plants throw the fertility back to the clay instead of from the clay to the plant.

The plants will build the fertility up in the soil and the plants will starve to death. Now you see what I mean when I had a different vision of plant nutrition in the soil than solubility. This thing is delicately balanced, but who has a vision of it. If you put chemicals into that soil you've ruined that cell root. These laws of physiology hold. It doesn't make much difference whether it is a person or a plant. I'm convinced that the Creator knew his business and man still hasn't learned.—*Dr. William A. Albrecht, in an interview with Acres U.S.A.*

CHAPTER 12

Proteins for Protection and Reproduction

PROTEIN SHORTAGES are intricately connected with the disturbed behaviors of animals, plants and microbes, all of which are successive sections in the pyramid that has man at its apex and the soil as the foundation of the whole structure. In the preceding discussions, the soil fertility pattern was shown to be developed by the climatic forces of rainfall and temperature. That pattern outlined one major region of healthy animals where more proteins are grown naturally in the crops, namely, the midcontinent of the United States. Then it outlined the region to its east and that to its west where in each, less proteins are grown in the crops and where purchased protein supplements are the problem for livestock production. Plants with higher and lower capacities to synthesize more complete proteins find themselves distributed in correlation with this soil fertility pattern too.

Man's migrations on the earth were possibly also determined by the soil pattern of fertility elements not only directly for his better foods and vegetables, but also for growing his animal proteins. Man and his livestock as his food are delineated by the same fertility pattern undergirding both of them as warm-blooded creatures of highly similar physiologies. Even for the control of man through the animal proteins he grows, the soil's pattern may be more subtle and more uncompromising than any politics, policies of colonization, or other politico-sociological forces. It is the soil that determines the proteins, not only by which all life forms are grown, but more significantly, by which there is given natural protection against diseases and by which reproduction of the species is regularly possible.

ONLY FERTILE SOILS GROW COMPLETE PROTEINS AS NATURAL GUARD AGAINST DISEASES

The provision of proteins by more fertile soils in any area does more to delineate the different life patterns and different health patterns of a single life form than almost any other ecological factor. It is the protein compounds of high quality alone that keep life flowing. They build the living part of the body tissue. In fact, only they represent growth as cell multiplication. Added fat may be added weight, but that is not growth of the kind just described. This stricter interpretation of the term "growth" is quite different, of course, from it when considered simply as so much "gain" (in bodyweight). Carbohydrates and not proteins have been the major constituent of feeds to give "gains." Weight increase has been the common concept of growth applied to animals in the pasture and feedlot. That is naturally so when the hanging on of fat and the loading of the tissues with water serve so well to make the practice of buying low and selling high a lucrative one.

But even then the success of this speculative venture demands the exclusion of the animal's reproductive potential. The feedlot phase of agriculture restricts itself largely to fattening the castrated males. Significantly, this practice finds itself located mainly on soils where the native crops serving as fattening feed grow bountifully as more bulk, but are so deficient in proteins—not only in totals but also in nutritional quality thereof—as to demand protein supplements imported from other more fertile soils or from places where plants providing more complete proteins are grown.

The growing of the younger animals to be fattened occurs on the soils much less developed under the scant rainfall of the western part of the midcontinental area where the animal grows itself "on the range." This leaves much territory for the young animal to clover. It allows the animal's growth of muscle meat according to its own choices of forages. Those are grown there on soils with much higher mineral contents. They are soils of many minerals well-mixed by wind action and much dust. The forages are composed of many legumes and protein-rich grasses in the scanty annual crop. By using much of the former buffalo territory farther west, the production of the beef calves for shipment to the east as feeders is more nearly a national biological performance in which the animals are managing their own production while the owner's part is mainly what an extensive cattle operator in Texas reported when he said, "We don't produce them, we only count them." It is almost a wildlife study.

The business of growing the beef cattle on the partially weathered, more sandy soils under scant rainfall means that such soils have not had their original fertility leached out by extra water percolating down through them. Their potential for growing quality protein in the vegetation, and thereby in the animals, is high, even though the total output of that per acre is low. The fattening of our beef cattle on the highly-weathered, humid soils which were developed excessively under rainfalls for much crop bulk per acre means that those soils have a low potential for growing quality proteins. They fail, thereby, in doing for the animals what proteins must do in addition to growing animals naturally, namely, protect them against disease and similar troubles while also guaranteeing fecund reproduction.

Such highly developed soils must, of necessity, contain more clay and little of the unweathered reserve minerals serving as sustaining fertility—rather than soluble starter fertilizers—to be weathered for plant nutritional service. We say their clay is acid, which merely means that it has long lost the exchangeable calcium, magnesium, potassium, and other fertility elements regularly restocking it as the result of the weathering of the reserve minerals for nutrition of crops. The plants have traded hydrogen, or acidity, from their roots to the clay to replace its fertility with acidity. The hydrogen is concentrated there as the "sold-out" sign of quality of protein in the crops grown on it.

While the business of moving beef calves from the west where they are grown to the east where they are fattened may seem to be an arrangement under our economic controls according to demands by the market, those activities occur according to nature's controls too deeply situated to be readily

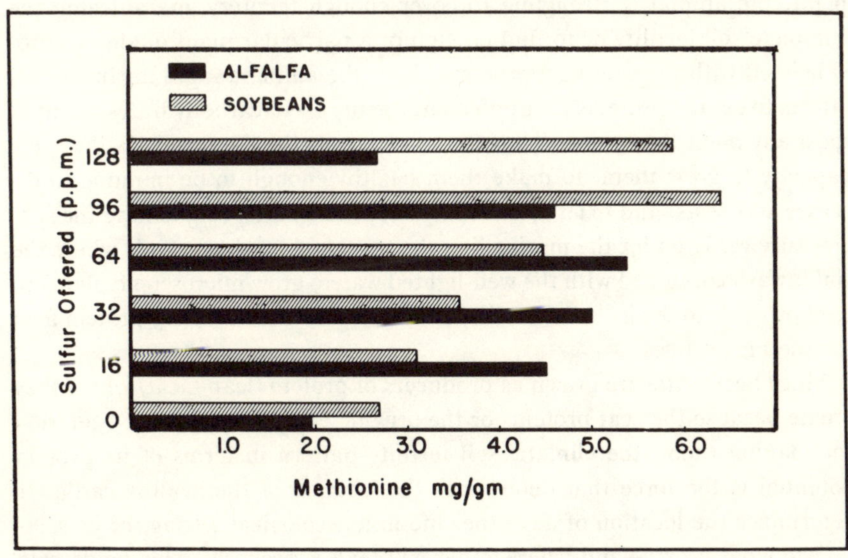

Alfalfa and soybean hays vary more than a 100% in their concentration of the amino acid methionine according as the soil has or has not a good supply of sulfur in its fertility store. Alfalfa hay goes to a higher concentration on soil treatment than the soybean hay.

recognized. Underneath the demands as economic forces, there is in reality the specific compulsion by the deficiencies in the proteins in the crops that go back to the soil fertility pattern.

Those deficiencies in potential for growing protein on humid, eastern soils are not necessarily remedied by putting into the vegetation more fertilizer nitrogen—the symbol of crude proteins of which nitrogen makes up 16%. Instead, this controlling deficiency is rather the shortage within the feed and food supply of some of the protein components, namely some of the amino acids—more commonly the tryptophane, the methionine, and the lysine. Commercially now, these are offering hope as uplift for the quality of the "crude" proteins we grow, if we can find sufficient supplies of these deficient amino acids by their separation from some suggestive products presently considered "wastes."

That the array of the different amino acids within the same plant species is not always in constant ratios is now well established. That this array shifts toward lower quality of protein for animals and man is some more commanding information from the careful studies of amino acids in crops in relation to varied soil fertility. Plants protect themselves and reproduce themselves only as the higher soil fertility supports their biosynthetic processes of converting carbohydrates into proteins by that help. Because the array of amino acids varies with the different species of plants and with the fertility growing any plant, the animal is struggling to cover enough territory guaranteeing the complete soil fertility or to find enough of a particular plant quality to provide itself with the complete proteins. Thus the animal ranges far in its own efforts to get the proteins to supplement the supply of carbohydrates found in most any plant that grows. Thus the soil controls the life forms via the soil's capacity to grow them, to make them healthy enough to be immune to diseases and pests, and to let them reproduce enough offspring to have the species survive. Even for the marine life, the sea supports that mainly where the soil inwash combined with the well-lighted waters grow microscopic plants to feed proteins to the little fish so they in turn can be the necessary protein feed for the bigger ones.

Since beef cattle are grown as producers of protein (lean meat) which they create because they eat proteins or the organic compounds in the vegetation for making them, the climatic soil fertility pattern in terms of its protein potential is the force that determines the location of the healthy cattle. It determines the location of any other life in its ecological setting, or its relation to what grows—and where—that will feed it properly. This holds true even for man on a larger geographic scale than for cattle, save as (a) his technologies give lifelines for his livestock and himself to bring in the fertility (or such in his food) to his more nearly local soil from distant more fertile ones,

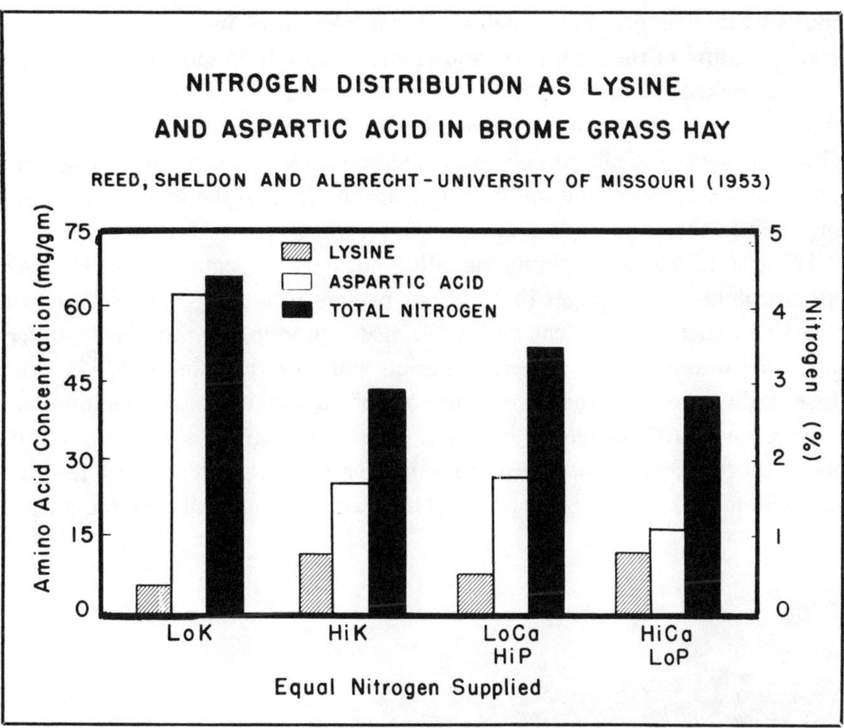

The concentration of amino acids (mg/gm) in the crop product, illustrated by lysine, does not necessarily follow the concentration of nitrogen (%). The latter may be high when the former is low, as illustrated by the low potassium (Lo K) and high potassium (Hi K). They may be also in reversed relation under low calcium and high phosphorus, (Lo Ca, Hi P), and high calcium and low phosphorus, (Hi Ca, Lo P), even when the amount of nitrogen supplied was the same in all cases.

or (b) periodically let the animals or himself make excursions out far enough and often enough to satisfy the hidden hungers and to guarantee protection and reproduction without which extinction must result.

NATURE GUARDS HEALTH THROUGH FEEDING, NOT THROUGH BREEDING

The soil areas favorable to growing healthy animals and healthy man, according to surveys of teeth, of histoplasmosis, and of other health irregularities, are those where the soil processes under the particular climatic forces of rainfall and temperature are breaking down the rocks and minerals to provide a flow of all the chemical elements essential for the plants. These essentials are adsorbed on the clay, and exchanged from there to the plant root for the hydrogen or acidity (a non-nutrient) it offers in trade. They must flow along that route in such amounts and in such ratios to each other as will nourish those plants, synthesizing the complete proteins. Such conditions prevail mainly in the temperate zone under rainfalls so moderate that dust may be picked up often by the wind and carried on for deposition under higher rainfall and more rapid weathering for services in plant nutrition. This represents the climato-chemical dynamics by which nature's assembly line of creation moves the soil fertility from the rock to the soil to the plants and then to the animals or man.

Those soil areas grow plants including not only the legumes, but also the other protein-rich herbages that put our protein-producing beef cattle (lean meat) and sheep (lean meat and wool, also a protein) on those soils under range conditions. These domestic animals will seek the same soils which in their virgin state made the brawn and bone of the wild buffalo, but supported no extensive cattle-fattening industry. It is those same soils where wheat makes more protein, too, unless the soils are already threatened with fertility exhaustion through heavy cropping. Those are the soils still carrying ample

Cattle in the Piney Woods, growing on the soils developed excessively under high rainfalls and temperatures, travel long distances through the woods to eat the herbaceous growths along the concrete highway. They graze the herbage most closely right along the edge of the concrete where root contact or transfer of calcium and other fertility from the cement through the soil suggest a higher concentration of protein in the herbage from this imported extra fertility. (Photo by courtesy of H.B. Vanderford)

lime or calcium, and by that same token are still retaining the other fertility elements which would have gone out had the higher rainfalls taken out the calcium to change them into acid soils or those too excessively developed and too deficient in fertility for fuller protein production. Under too little rainfall for big yields of much vegetative bulk per acre, they are the soils where the cow went ahead of the plow while assaying them with a favorable report for her good health and for fecund reproduction of herself and likewise of man. As man pushes himself and his livestock off these protein-producing soils on to the fringe soils, he must extend the lifelines from the latter back to the former, except as he and his supporting herds and flocks can tolerate more of what is called diseases and more of malnutrition and partial starvation.

Much reliance is put on the belief that by selecting and propagating certain plants of a crop, we can eventually find those which will tolerate diseases like smut, rust, root-rot and others. Much is said about breeding resistant crops or those which will tolerate such troubles. We fail to see the germ diseases as attacks by those invading foreign proteins, in their struggle to get their necessary protein while parasitically taking over organically elaborated materials of that kind as starter compounds from which to synthesize their own. We fail to see that "immune" plants are those getting enough soil fer-

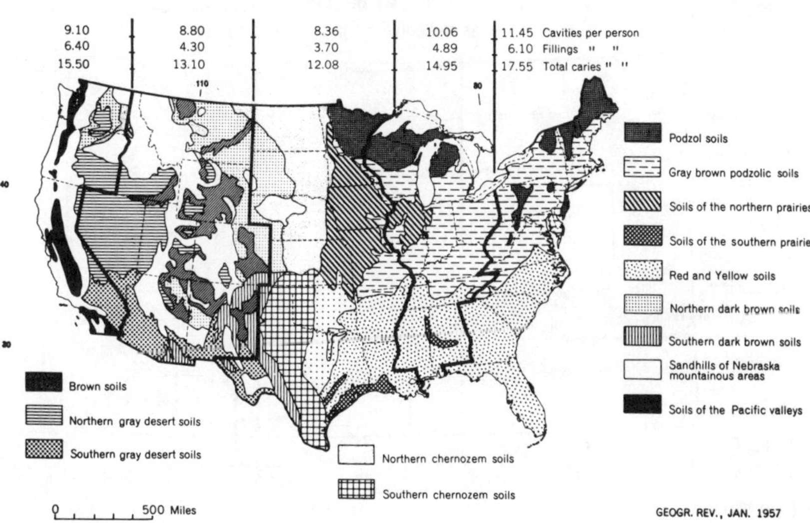

The number of cavities, fillings and caries (cavities and fillings combined) in the teeth per person of nearly 70,000 inductees in the Navy, reflect the teeth condition of young men in correlation with the pattern of soil fertility. We are slowly connecting not only animal health with the soil fertility, but human health also.

tility support for creating their own protective proteins or antibiotics in the same way as fungi make theirs to protect themselves from each other, and to protect us similarly when we take their antibiotics into our bloodstream.

Any hope that we might "breed plants to tolerate diseases" is a vain hope when it was neither drugs nor poisons but soil fertility (both organic and inorganic) which protected the virgin crops grown into their much-admired ecological climaxes of pure stands of nearly "perfect" plants. If deficient plant nutrition, especially with reference to proteins, brings on disease and pests as nature testifies and has been experimentally demonstrated, then to believe that we could breed for such resistance is the equivalent of believing that we could breed a plant to tolerate starvation. An experiment set up to test this hypothesis would last only one generation and would be no more logical than breeding to establish a race of bachelors. A very essential but missing part would ruin the hopes of the anticipated or planned results from the experiment.

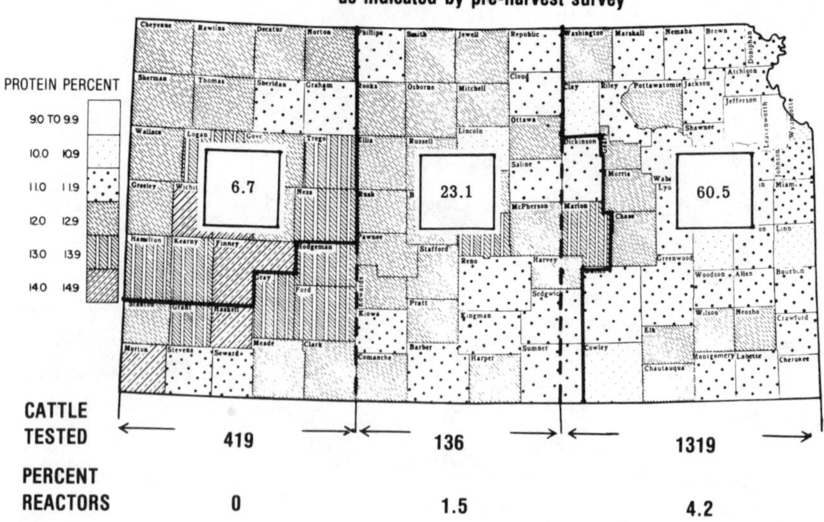

The number of students native to Kansas by areas gave less positive reactions to the histoplasmin test, or had less histoplasmosis disease, as the soils were more fertile in terms of the higher concentration of protein in the wheat by county averages of 1949. (Percentage figures inside squares.) Cattle have had less histoplasmosis according to the same pattern (percent reactors).

We fail to see that each form of past life, natural or domesticated, must have protected itself before we became concerned about its failing health under our interference with its natural biotic behaviors, to make it suit our demands. We have not appreciated what those natural behaviors contributed by the animal instincts will do if we cooperate with them for growing healthy animals protecting themselves by means of good nutrition, including plenty of proteins. We hold postmortems, tabulate symptoms, and offer explanations (often only consolation) but fail to comprehend causes. We run the motion picture backwards, as it were, in our delusion that we manage the lives of animals and that we control the ecological pattern. We fail to recognize the fact that the soil fertility is the basic force under all creation.

Thus, there is generated a blind faith that with our many technologies, we can extend our livestock (and ourselves) over vast acreage of previously unused lands. There is little or no thought given to the reasons why those have so long remained unused by living creatures. History makes little use of its records which are quietly telling us that some animal, or some man, has already tried such areas with the resulting failure to maintain itself or its kind here. Such lands remain as acres, but the serious shortages in the soil's possible chemical dynamics for growing the proteins are reasons for absence of living things. Those are the hidden hazards to be searched out before taking those acres over for agricultural production. Such acres may appear as tempting places to try a grass agriculture and livestock. While the seriousness of many hazards to animal production may be subject to debate under economic and population pressures, no one has yet been willing to debate the negative side of the simple proposition that "Animals (and man) must eat, and particularly of the protein," if they are to enjoy protection from diseases or pests and if they are to multiply their own kind and their species are to survive.

MICROBES, PLANTS AND ANIMALS CAN PROTECT THEMSELVES BY PROTEINS OF THEIR OWN CREATION

When we use antibiotics, that simple act represents the acceptance by us of the synthetic services from the lowly microbe for protection of ourselves through what the microbe grows within itself for its own protection. From down next to the soil at the bottom of the biotic pyramid, these biochemical services, approaching those represented in the synthesis of proteins, are passed up to the top of that structure for our protection against other lowly but dangerous microbes. Plants, too, offer us protection in the many compounds simulating proteins when they give us vitamins, hormones, and catalytic or stimulative effects still unknown but ascribed to the so-called protective foods. Proteins and compounds so near like them are the major protection against invasions of our body by foreign proteins and against degeneration of most of the warm-blooded bodies. These foods give us their protection only when the fertile soils have grown them for that capacity.

It is in terms of specific protein compounds lending themselves to commerce that our animals give other animals and us protection when we use serums, vaccines and various inoculants made from animal blood proteins. It is the cow that can take even our disease of smallpox, and live through the scourge of it; can build proteins in her bloodstream to protect herself against recurrence of it; and can then share those protective proteins with us for our vaccination and immunization of hundreds of humans. Yet she does that clever physiological stunt by support of no specific drugs. She does it by the support of nutritional compounds no more startling than those in green grass growing on fertile soil.

The horse takes our form of typhoid fever. With apparently no disruption of his health, he creates proteins which combat the effects of the typhoid bacteria and share those disease-fighting compounds with us as our guards against those foreign proteins. He does this by his own inoculative protection against such bacteria. The bodies of our animals, and our own bodies likewise, often suffer from insufficient proteins and protein-like substances for protection against the invasions by foreign, death-dealing, germ-proteins. Yet with a little help brought to us by the microbes, the plants and the animals, our bodies are seemingly stimulated into the specific activity of creating sufficient of our own proteins for protection in what we call immunity or antibodies.

That plants protect themselves from fungus troubles by means of proteins has previously been illustrated. That even non-legumes resist attacks by insects more completely as more nitrogen and more lime in the soil suggests, their growing more specific protective proteins has also been demonstrated. If these are the facts about nature's controls of such troubles, they challenge our acceptance of the truth of the converse, namely, that the increasing fungus diseases of our crops and the mounting numbers of insect pests on them are premised on the deficiencies of protective proteins in the plants and of those, in turn, on the soil's deficiencies or imbalances of the fertility elements and organic substances for growing them within the selected crops.

For the making of the antibodies as protectors in our own bloodstream, the lymph node system plays the major role. Very recent research has demonstrated the possibility of building up the body's capacity to make more protein antibodies by transplanting parts of lymph nodes into the body tissue, with effects much as those from inoculations into the bloodstream. That tissue transplants build more antibodies has been demonstrated for good results in antibody productions against typhoid fever. According as the body was initially weaker in making antibodies, the tolerance of the transplant was reported higher. Here is another case of putting into the tissue—not into the digestive system—one kind of protein serving to stimulate growth of specific protective proteins. But these protectives cannot be grown by a body except by nutrition with proteins.

Shall we not open our minds, then, to the possiblity that the shortages in proteins, and all that is associated with them in their synthesis by microbes, by plants and by animals, are keeping our livestock (and us also) under a constant kind of malnutrition and are prohibiting all warm-blooded bodies from collecting and creating the necessary proteins by which such bodies can protect themselves more completely or build their own immunity as their progenitors must have done before domestication?

Seemingly our wild animals gather their own "medicines" by instinctive selections, among different plants and among the same plants growing on different soils. Our domestic animals suggest their making similar selections within the limits permitted by our enclosure of them with fences, barns, stanchions and other hindrances to their free exercise of choices for their own better nutrition and better protection against diseases. There is the suggestion that our pasture management, which uses three or four single crops in sequence during the season for the cow's harvesting services, is militating seriously against her chances to graze herself into good health as she would in taking mixtures of many plant species in her herbage. Perhaps our management of pastures is economically successful as a system but the cow is having difficulty in living through the operation of it.

It was an older scholar of animal nutrition, struggling to produce feeds for better health of animals, who had originally practiced veterinary medicine and who reported: "I had held postmortems of many of our domestic animals in trouble with many kinds of ailments. I thought I was familiar with the internals of most any body. But it was not until I went deer hunting in the mountains, bagged one, and opened it in preparation for transport home, that I got a clear view of what the internals of a truly healthy animal really look like." Have we been feeding animals on deficient soils so long that we have no vision of what a fertile soil can do for growing healthy animals, or even what they can do for their health by their own choice when they graze over more different soils and different plants?

REPRODUCTION DEPENDS ON THE BIOTIC STREAM OF PROPER PROTEINS

It was some experiments, using sheep as farm livestock and domesticated rabbits under carefully controlled procedures, which demonstrated forcefully the fact that soil via the protein output in the feed crops controls the reproduction possibilities. Ewe lambs were fed on legume hays grown on less productive soil given (a) no treatment, (b) phosphate, and (c) both limestone and phosphate. Their growths as increases in bodyweights were in the relation of 8, 14, and 18 pounds per animal per 63 days for the above soil treatments, respectively, when equal amounts of hay per head per day were consumed. Differences in the wool were soon visible during the feeding period. That from the lambs fed the hay grown on the soils with more complete treatment (limestone and phosphate) was the only one among the three lots which could be scoured and carded without the destruction of the fibers. How differently they would have taken dyes if the fabric was not tested. More significant, however, than the failure of the skin to secrete this healthy protein fiber in the case of the two lots of lambs fed on the hay grown on soil given (a) only phosphate and (b) no treatment, was the failure of sheep in reproduction as a consequence of failure to uplift the fertility of this soil growing their feed.

When at the age of eighteen months the three lots of lambs were put with the ram, these two lots producing poor wool also failed to give a lamb crop. The third lot, fed previously on hays grown on soils given both limestone and phosphate for soil fertility improvement, gave a lamb crop as the result of the mating with the same male.

As an additional test of the level of fertility as the cause connecting the soils, the proteins, and the reproduction, the two hays grown on the same soil treatments, (a) of phosphate only, and (b) of both limestone and phosphate, were fed to two lots of male rabbits in use for artificial insemination. Their regular delivery of semen was measured carefully and studied critically, only to find (a) the delivered volume decreasing, (b) the concentration of spermatazoa falling and (c) the percentage of live spermatazoa declining rapidly for those rabbits fed the hay grown on the soil of which the fertility was improved by no more than only a phosphate treatment.

Such was not the case, however, for those rabbits feeding on the hays grown on the soil given both limestone and phosphate. No significant irregularities in the production and delivery of the semen were manifested by this second lot.

When these differences between those two lots of rabbits as potential procreators had become especially wide, the males of the former lot were approaching sexual impotency so closely that they were indifferent to the presence of a female in oestrus. At the same time, those in the latter lot manifested their interest in her the moment she was brought near their hutches.

Still more significant, as evidence of the relation of soil fertility and crop proteins to reproduction, were the marked changes in reproductive potentials resulting when the feeding program of the rabbits was modified merely by interchanging the hays for the two lots of rabbits. Only three weeks had elapsed after this shift in feeding when the lot of originally impotent and indifferent males was restored to sexual vigor with all the characteristics of potent males. The formerly potent ones exhibited falling curves for all of the measurements. In the same short period of three weeks, those on the hay grown with the limited soil treatment had fallen to the same low level of the other lot before the hays were interchanged.

When, in these tests, the soil treatments for improved production of protein by legumes, as measured in terms of increased nitrogen in their hays, were the only variables responsible for shifting the sexual vigor from impotence to potence or vice versa, one can scarcely refute the causal connection between soil fertility, proteins, protection and reproduction. It is becoming more clear that the complete proteins as food compounds are connecting the animal (a) in its survival as an individual via nutrition and protection against disease and (b) in its survival as a species via fecund reproduction, very definitely with the required combinations of the nutrient elements offered for plant growth by the soil.

OUR PERVERSION OF THE STREAM OF LIFE

We need not strain the logic of our thinking to connect reproduction by our livestock with the proteins grown by the soil as their feed, if we remind ourselves that: (a) proteins are the only compounds carrying life, (b) reproduction is the transmission of life in units of the single cell and (c) the characters of each body are carried in genes as parts of the chromosomes—all of which are living proteins. Any reproducing cell splits or divides its chromosomes or sets of genes. Then, after dividing, each of these chromosomes as sets of genes, must grow back to normal size again before they can carry on the next in the succession of divisions. If as split units, or reduced masses of protein, they are to regrow, they must be given protein via nutrition for that restoration after each division. Should that supply of nutritional protein fail in even one of the required amino acids, then some inheritable property or process of the cell will not be transmitted, but will be lost and missing from the succeeding generations. That loss need not kill off the species but may be a contribution to the degeneration of it or reduce its chances for survival. Such a deficiency may bring a change sufficient to give what breeders call a mutant; consequently, deficiencies of the amino acids in the nutrition of this reproductive process may be a mutagen or a generator of mutations. Drug-like organic compounds are increasing by which, as treatments, changes are brought about in the genes or mutations amounting to deficiencies in the synthesis of several of the amino acids. We are thus seemingly changing the characters of the animals in successive generations by deficiencies in proteins. Those changes are losses, not additions, in the chances for better health and better survival of future generations.

Proteins for Protection and Reproduction   161

Shall we not raise the question, then, whether in our emphasis on fattening under the minimum feeding of even the crude protein only, such protein deficiencies would not have meant losses in the genes or losses of properties and processes with successive generations? Might we not have been dropping properties out via changes in the genes until the property of vigorous body growth has become the deficient growth which we call the "dwarf?" We certainly have not been selecting and propagating cattle for vigor in survival of the species. Hence the dwarf might well be viewed as the result of our perversion of the life stream toward fattening, with its failing physiologies, rather than towards the survival of the species as was the direction of the life stream under evolution before domestication. Shall we not view deficiencies in the soil under our exploitive use of it, then, as they may be bringing about dwarfism as a testimony of slow extinction of the very livestock we should direct toward better survival by better nutrition through more fertile soils? Will the life stream continue to flow, in spite of us, if dwarfism is testimony of changing generations under protein deficiencies in the reproductive processes residing in and carried on by only proteins?

If we see in the cow, or in the bull, some anatomical or even physiological symptoms of what might be the accumulated deficiencies of nutrition, culminating in the stream of transmitted life running so low that the fetus, grown up to the time of parturition, is only a small replica of what it would normally be, we might be inclined to believe that some new gene has been added and is responsible for transmitting this strange character of growth failure and impending extinction of the species. Surely we would not expect this character to show up in the next generation when there might be no second generation. It seems more logical, and in better accord with facts of evolution, to believe that in consequence of our selection and propagation under economic pressures for cheap gains, the protein stream of reproduction and inheritance would have far greater chances to suffer losses from what it has formerly been transmitting, then it would have to make additions to such. Genetic studies have been legion in which there were knocked out by X-rays, ultraviolet rays, triazine, nitrogen-mustard and other chemicals, some properties formerly transmitted but missing forever after in the succeeding generations.

In accordance with this latter more common occurrence of changes in the offspring and its decreased reproductive potentials, it seems more logical to see this failure, of a midget calf to grow, as one of losses or of several deficiencies in the biochemical processes centered in the proteins of reproduction by which they fail to give what we consider normal. The whole trouble appears as deficiencies or losses and not transmission of additions or something new. In that respect it may well be considered the expectable when we see declining protein in the succession of crops and failing reproduction in what the farmer sees when he says, "I must get some new variety of oats; the kind I have is running out." Is he not merely reporting that the species is failing in nutrition equal to reproduction of itself or continued survival on the soil growing it?

If we should try to establish some visible symptom as proof of dwarfism by a blood test of it, and if it is a deficiency in the qualities of the proteins built by the dam or sire and transmitted in the inheritance stream, should we not expect an animal suffering from this deficiency (positive to the test) to demonstrate low protein activity in its blood in response to most any injected foreign protein or even protein-like substance? If the healthy animal has been building ample proteins for generations with self protection and fecund normal reproduction resulting regularly, should we not expect its blood reaction to a foreign protein or test substance to show up quickly in the extra defense proteins in the bloodstream to match the foreign ones with their digestion, agglutination or other opposition effects soon resulting? Self protection against foreign proteins in the blood is the mark of health via the body's good nutrition to build ample protective proteins. Such test itself, when positive by slow defense reaction points to the deficiency in proteins for reproductive processes and transmissions going back to nutrition and even to the soil fertility, should not seem so abnormal or so unusual as the precursor or even the cause of dwarfism.

While it is consoling to establish validity of the symptom, research is far more fruitful when it makes attempts at prevention and other aspects of health management, or when we establish the causes rather than only label the trouble with catalogued symptoms. Such is well illustrated in the case of malignancy of the human body or even for a headache. But in terms of managing the consoled situation, we are not capable of doing anything but to cover the symptom with a relief treatment often more damaging and distressing than the toleration of the unmitigated symptom. But when the trouble may be due to a deficiency, as once were many baffling diseases with particular symptoms and causes unknown, there is little managerial possibility until the deficiency is specifically located. Dwarfism suggests itself as a case well viewed as a deficiency, pointing toward a protein deficiency or a deficiency of something connected with protein processes.

If blood compositions are studied in this light, the differences in the chemistry of blood between the afflicted and of the healthy animals might give helpful suggestions. The numerous cases of earlier degenerative diseases of the individual, later discovered due to deficiencies, ought to turn our thinking toward prevention and toward saving the afflicted, rather than condemning them to slaughter as an escape from the problem. When we recall the many diseases like kwashiorkor, a deficiency of protein; beri beri, one of the thiamine (vitamin); perosis, of manganese; cleft palate, of riboflavin; pining or coast disease, of cobalt; parakeratosis, of zinc; anemia, of copper or iron or both; demyelination of the spinal cord, of copper; and many others, should we not look at the failing biology of agriculture as caused by deficiencies going back to the soil fertility and possibly as our sin of omission rather than as the sin of commission by the animal for its condemnation to death?

When plants get their proteins in varying degrees of completeness for their reproduction by seed, according as the more complete suites of fertility elements in the soil permit; when herbivorous animals must depend on plants for their proteins as a collection of all the required amino acids; when protection against invasion of warm-blooded bodies by death dealing agencies is given by proteins; and when the stream of reproduction of any life can be kept flowing only by means of proteins; shall we envision the possibility of any life form stepping outside of this pattern of constructive controls? Can man by his technologies put his animals beyond them even with his extended life lines now already becoming tangled with life lines of economics and politics? Only slowly will the science of the soil become our knowledge common enough to bring the realization that the chemical performances in the soil connected with plants and animals are the sources of biotic agricultural health first and national wealth second. Only by fertile soil can there be created ample supplies of the complete proteins by which not only growth but protection as health, and reproduction as survival of the species are possible.

CHAPTER 13

Impoverished Soils, Poor Animal Health, and Distorted Economics for Agriculture

THE COSTS OF GROWING healthy livestock and healthy people do not fit themselves readily into our economics where costs and earnings must always be matched in monetary values (dollars) lending themselves to subtraction, divisions, equations and summations in only arithmetical terms. We are slow to realize that good health is not a purchasable commodity. Its value transcends any price mark. We have lost sight of the fact that it must be grown into the body. It cannot be thrown in from purchases at the drug store nor introduced effectively as tissue implantations, or hypodermic and intramuscular injections by the physician's needle and syringe. In handling livestock, we can throw the fattening performance of the desexed male into high gear. But for the growth of a healthy body and its maintenance at that level, a guarded, steady pace for endurance and distance is demanded.

Our increase of wholly monetary views of agriculture, as if it were only an industry under economic pressures which consider all the costs and all the returns from it only in monetary values, have been—in no small sense—prohibiting healthy livestock. We need to take a new view of some phases of the farming business, which do not balance out so readily in economic equations, to be solved by no other factors than monetary units. Health of plants, of livestock and of humans via proper nutrition is just such a phase which will not submit to solution by monetary manipulations. We need to shift away from the belief that all agricultural problems can be solved by such procedures under emphasis of the economics only.

SOILS EXPLOITED OF THEIR FERTILITY CANNOT SUPPORT PASTORAL FARMING WITH HEALTHY COWS

We have been prone to ridicule the simpler arts of agriculture in the older countries and in the older civilizations where the plow and other modern agricultural machinery followed, rather than preceded, the cow. Just now we are engaged—on an almost international scale by dollar methods—in supposedly educational activities of a missionary nature. Those are aimed to bring these ancient agricultures up-to-date in the economics of applied agricultural mechanics for mining their soil fertility faster. We are unmindful of the fact that in these older countries, the cow was their nutritional chemist; she searched out the soils and her masters capitalized on her judgment of them as the soils fitted to be forever under a highly pastoral agriculture of healthy animals. Those soils were not first exploited by a highly arable agriculture, as we know it here, if the European manure pile in the front yard or the tank wagon, flowing its liquid manure on the pastures and meadows (to the American tourists' disdain and disgust), dare to be considered as reliable indicators. Had the older agricultures mined their soils at the rate we do ours, they would not have survived long enough to become old.

For us, the plow has been the emblem of agriculture. But it has always been ahead of the cow. This was not so unexpectable in the age of developing much farm machinery, of more internal combustion engines, of the mining, conversion and combustion of fossil fuels, of labor-saving devices, and of rising standards of living. It was the most expectable on soils with great stores of virgin organic and reserve inorganic fertility. Our midcontinental soil, especially, were of most extensive areas, very level topography, silty texture coming with windblown origins, high fertility in exchangeable forms on the clay, and rich in ready reserves of nutrients in windblown minerals brought in wide varieties from the arid west. Such soils naturally invited the plow and other mechanically exploitive means. Soil conditions of this type are natural temptations to convert them into cash crops and not to consider them for their creative fertility capital for growing living things. They are taken as a quick means of mining them and collecting monetary capital for coupon clipping service under self-perpetuation ever after. Surely such land would invite the view of agriculture as another case of banking operations with mounting capital under guaranteed returns. This is far removed from growing plants and animals and the daily routine of caring for all of these to insure their healthy growth and fecund reproduction.

Such a view of the soil in only its potential to build up monetary accumulations lets us move the cow to those soils which are either not sufficiently, or already excessively, developed under the climatic forces for naturally nutritious feed production for healthy livestock. It puts the cow under handicap for her own good health. Now that (a) the seriousness of erosion is being recognized, (b) no more areas of fertile soils are left to be so easily exploited, (c) the fertility decline is becoming apparent after being hidden so long under crop juggling for only yields of bulk, (d) the problem of protein supplements as animal feed and the health troubles in animal production are being accepted as malnutrition via deficiencies traced back to the soil and

not alone to the feed store and the veterinarian and (e) we are saddled with
the responsibility of being Santa Claus for a much more inflated and hun-
grier world, we are talking about less plow but more cow from a grass agriculture that will cover any soil and give us less fat but more meat and more
milk. We moved the cow all over the country. But we forgot to build up the
soil fertility in each locality to the same high standard which the buffalo
demanded for his survival. The buffalo standard of fertility of the soil represents also a better soil for the cow, when naturally fitted to grow her, than is
any we can yet fertilize in trying to mimic buffalo soil to guarantee animal
health by chemical soil treatments.

While all these problems are readily ascribed to irregularities in mainly
economic and social arrangements, we are reluctantly coming to recognize
the low, the declining, and the imbalanced fertility of the soil beneath the
whole structure. We, ourselves, led the cow to soils contrary to her choice of
the fertility there. We failed to see the fertility pattern according to different
degrees of development under the climatic forces as the corresponding and
controlling pattern of livestock health. Our fences confine the poor beasts to
the deficient soils in the pastures which are growing worthless weeds in place
of nutritious herbages. In similar manner, our technologies have extended
agriculture over less fertile soils by its many forms of so-called crop specializations which are in reality exhibitions of cropping limitations. The possible
crops are limited because the soil fertility is not economically modifiable for
plants requiring higher physiological potentials. Such specializations like
cotton farming, sugarcane culture, forest farming, and others are illustrations of what occurs on soils of which the fertility would not entice the cow,
and of which her assay would declare them too deficient to support her with
good nourishment.

Do you suppose the cow would select a forest site, clear it, and expect the
planted crops to be good nutrition when the Creator himself could grow only
wood there, and then that only by the return annually to the soil of all the fertility in the leaves or needles? We have allowed the simpler mechanics of
growing any kind of grass and the more favored economics of letting the cow
graze it to dominate our thinking of economics so completely that the physiology of the plants and the physiology of the cow eating them have been given
little consideration. By way of the deficient soil fertility, we must be reminded
that the cow is more than just economics or than just a mowing machine or
just a hay baler. Her body's physiological functions are not just economics
and mechanics. They are connected with, and dependent on, the fertility of
the soil. She is not asking for merely tons of forage and acres of grazing. She
is calling for complete nutrition to undergird the reproduction of herself and
establishment of subsequent milk flow of higher nutritional values for the

calf. She is not aiming to establish records of gallons of milky liquid, pounds of butter or hundredweights of over fatted carcass for cheap economic gains.

On these fringe soils, or those outside of the midcontinent where the climatic pattern of soil development located higher protein production in the forages, the grains, and the ruminant quadrupeds like the earlier bison and the later beef cattle, we cannot expect the same animal health without soil treatments. This expectancy drops still lower and rapidly when we have farmed the area as land more than the soil as nutrition for livestock. Exploited soils will not support a pastoral farming with healthy animals as the output.

LAND MAY HAVE ONLY SITE VALUE
SOILS MUST HAVE PRODUCTIVE VALUE

In our concepts of the economics of agriculture, the land area has become fixed as property values and capital investment. But such capital is not self-perpetuating as is true of the mortgages on it or of bonds, stocks, etc. with rates of earnings fixed accordingly. Instead of the land area giving the earnings in the form of healthy plants and animals, it is the soil which must guarantee that, but under a mining rather than a managing operation. Land, then, in the prevailing agricultural economics becomes soil, and soil becomes consumable goods like the oil in the earth or the coal and other minerals in the mines. Agricultural production does not consider losses in declining fertility, nor the costs of restoring and maintaining it as items to be charged in with a price to be set on the grain or livestock to be delivered to the consumer if agriculture is to perpetuate its investments. On the converse, it liquidates the soil fertility by installments. It throws in with each sale a portion of the creative capital for living products which guarantee animal and human health without charging replacement costs in the products sold.

Yet our distorted economics of agriculture call that liquidation a case of taking a profit. Income taxes are charged against that expenditure of the monetary capital originally put into the purchase of the land. Taxes are charged even against the unearned but increased values assigned through assessment for taxation purposes by others neither managing the farm nor recognizing the soil fertility as the earning power. In that kind of economics for capital in land, we forget that land in its proper definition is dry surface area, or dry footing. It has but two dimensions, namely length and breadth. It is a location for agricultural business. Yet that business gets its earnings from the living processes in the growing of crops and animals. Taxes should not be charged on land according to these earnings annual, but on the fixed site value. Unfortunately, land tax moves itself upward on the land as the capital, while the soil fertility as the earning power, via quality feed for healthy animals, is on the decline because the fertility is gradually mined out under a liquidation that does not charge replacement costs for it in the goods sold. Yet this kind of agricultural business bears mounting real estate taxes to help pay for our higher standard of living which the community enjoys.

The soil fertility is the earning capacity of the land. The soil earns through the feed or food it creates in crop growth. Food and feed are values not classifiable, thereby, as such a simple matter as are dollars fitted into the economic scheme. Dollars alone don't guarantee or replace good animal health, as droughts illustrate so pathetically. Agricultural earnings must first be a biotic performance by crops and livestock. When these are harvested, marketed, and slaughtered, then, by their death they become commodities and take on the same dollar values of the economic pattern in any other business involving the sale or transfer of non-living goods. Healthy animals are the result of the biotic performance of the fertile soil. Its dynamics demand decomposition of its own mineral and organic matter contents. They call for

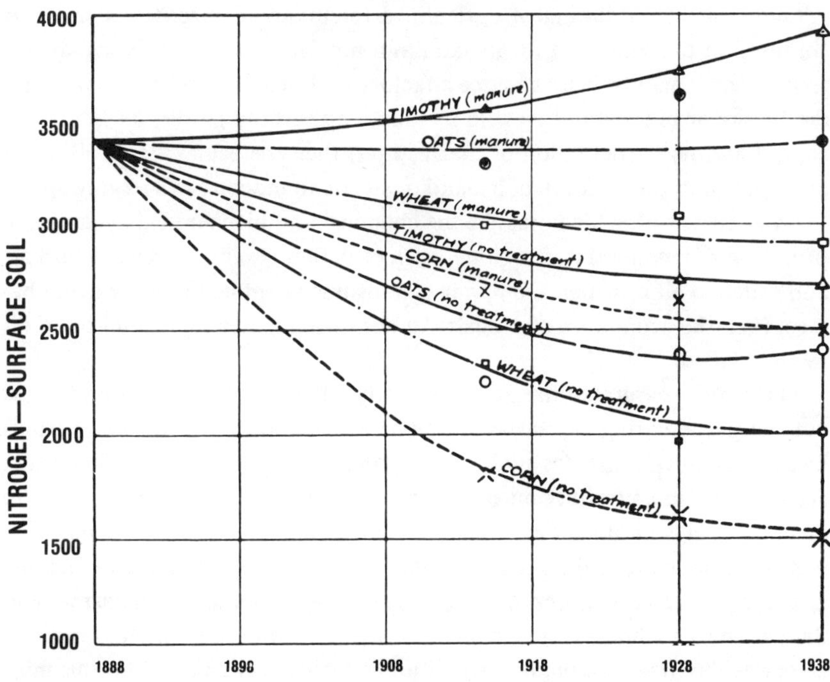

The total shock of nitrogen in our surface soils has been going down while the organic matter there was spent without its restoration. This decline in one generation under corn with no soil treatment (fifty years), put the nitrogen down to but one-half the original amount in the virgin soil. Soils so low in nitrogen mean poor chances for growing protein-rich feeds required for healthy animals.

the removal of soil fertility. But this removal is not viewed as installments in the liquidation of the initial capital. Instead, the economic view of the land takes the soil to be perpetually functional and never-failing in this creative production.

On the contrary, soils wear out; their fertility becomes exhausted. They must be viewed economically in their initial investment just as we view any other mineral mining business. Soils have not only the two dimensions of land, *i.e.,* length and breadth, but they must have four dimensions, namely depth and fertility. Ownership of farm property is, by title to the land, by its two dimensions which are fixed and permanent. Taxation of real estate is by the same characterizations. But the earning power of the land is according to the depth and the fertility of the soil. These represent the capacity of the soil to store water, to provide nutrient essentials, to recover productive capacity on resting from cropping, and to carry sustaining fertility in the reserve of unweathered minerals.

Our economic thinking of agriculture for taxation purposes is seriously confused in thinking of it for its land and not for its soil. Viewed as land in terms of only area or just as we see a factory building, and not as the industry on the site or in the building with its monetary earning power, its security of capital and its perpetuation of that capital, then the economics evolved for and applied in the taxation of its land or site of an industry will readily apply. But viewed as soil, which must be the chemical and biochemical dynamics of providing raw mineral and organic matters so they can be connected with air and water to all of which the power of sunshine is applied to grow living beings, then the economics of industry in non-living materials cannot be so directly applied.

The fertility capital of the soil, as initially taken over in virgin condition, is not self-perpetuating capital as are monetary investments with their various insurances and guarantees against losses and forced liquidations. Chemically active as the soil's fertility must be for creative production of living forms to be the earnings on the monetary investment in both the soils under management and the land as the site and economic unit of the community, that fertility removal in crop after crop means dwindling resources to the farmer and the nation as a whole. We have failed to see the distorted economics in those facts when we have not made corresponding taxation allowances for this mining of mineral fertility in agriculture, yet make corresponding allowance as high as 27½% in connection with mining the fossil fuels taken from oil wells. Economics of industry make industrial capital self-perpetuating. That kind of economics makes agricultural capital self-liquidating in a generation.

These distortions of our economics for agriculture are those evolved for industry which merely transforms non-living materials. That kind of economics then superimposed on agriculture will not fit in terms of sound economics and will not classify agriculture as an industry. Agriculture is but our taking over nature's creation of living forms according as the soil under climatic forces limits them. Our attempt to put a biotic performance by nature under industrial economics is the real reason for what is said to be the agricultural problem. Surely the economic standard of income, in monetary terms of agriculture, cannot be equated against that of industry when the two represent values differing so widely that they cannot be equated in similar dollar units. Industry builds up, multiplies, and perpetuates its earning capital as units of dollars. Agriculture spends its soil fertility capital in earnings in the form of crops and livestock which are consumables as food. Thereby, the capital invested in soil is not self-perpetuating by its earnings. It is fertility being liquidated to feed all the folks of the nation, a fact not yet clear in our distorted economics of industrial and business origin superimposed upon agriculture. Nationally we have been liquidating our food resources, but have viewed it as no more serious than a problem to be passed to the secretary of agriculture for his submission to the members of Congress for their economic adjustment through legislative appropriations from the national tax fed treasury.

BUYING ON A SELLER'S MARKET, AND SELLING ON A BUYER'S MARKET COLLECTS NO FUNDS FOR SOIL-SAVING

Somehow or other, we usually stumble into knowledge of natural matters mainly by postmortems, and seldom by prophecies and predictions built on basic natural laws. But thus it has ever been and is recorded in the ancient literature which reported that, "No prophet is without honor save in his own land." Even the first national activities aiming to do something about saving the soil from its suddenly increased erosion were initiated because of postmortems on the soils, either of depleted fertility or already gone out of production. From many postmortems, it dawned on us that the soil fertility is already too low when the land doesn't grow crops quickly enough to cover and protect itself from erosion, where it formerly was not eroding rapidly but now has suddenly washed away. We discovered that well granulated, fertile, surface soil full of organic matter does not erode rapidly as does the infertile, plastic subsoil of high clay content below it once the water has cut through to the latter and is running over that readily dispersible, infertile subsoil.

It took much publicity about the evils of soil erosion to wring money out of the federal treasury and set up a national soil-saving agricultural section in Washington, D.C. for the uplift of the farmer's efforts in saving his soil. That national soil-saving activity was a matter of much tillage mechanics to fight water running downhill, as such a liquid mineral always does naturally. We gradually came around to see the infertile, naked and broken soil-body inviting itself to be churched into slush by the impact of even the little raindrop. Emphasis was put on terraces as defense against moving water, when terraces are a kind of splint holding a broken limb of the body in place while the body as a whole is nursed back into good health and the injured part healed over. No one likes to farm over terraces forever any more than we expect to hobble about with splints on a broken leg forever. It took a few years of soil conservation activity to shift the emphasis—by a shift in name—from *erosion service* to *conservation service* in the national campaign for mechanical saving of soils with all the extra educational efforts under both national and state planning. Even now, apparently, the search for purpose and procedures, perhaps under still more appropriate terminology and classification, is continuing in the hope that the nation can do something to save the nation's soils.

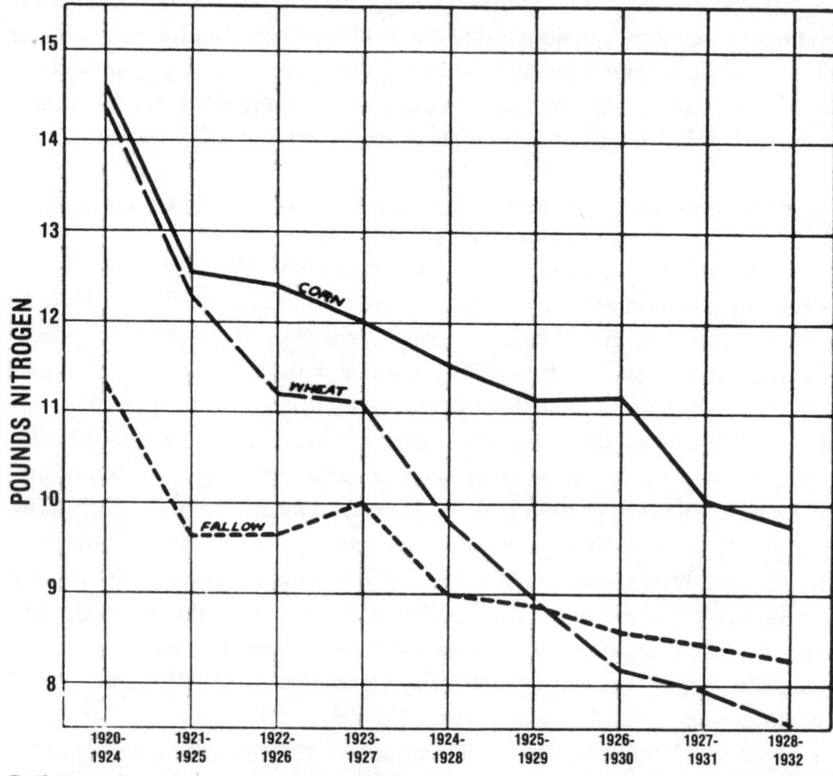

Soil organic matter under decay liberates ammonia nitrogen to be held by the soil, but changes later to nitrate nitrogen not so held and active for plant nutrition. Some sod broken out under measures of the seasonal levels of nitrate nitrogen reports the serious decline of this soil activity by advancing five-year averages where no soil treatments attempted to restore the soil organic matter.

That line of thinking, however, forgets that no one knows soils on each farm in greater detail than the individual farmer who is on his farm. Conservation on a national scale is a hope that the farmer who, in the last analysis must save his own soil, can be summed up with all other farmers through national encouragement to save the nation's soils. That fails to realize that the farmer has already done much soil-saving and can do no more than he now does, with his business in the economic set-up where he must liquidate his fertility capital to earn his food and enjoy the privilege and distinction of being a farmer, under national leadership moving more and more to manage his business for him.

In conserving the soil, the farmer appreciates fully the need for maintaining both the inorganic and the organic soil fertility as the creative capital. He has done well to purchase and return a little starter fertilizer. In addition to that fertility uplift, on many soils he struggles to rebuild and improve the sustaining fertility with limestone applied for its calcium and magnesium, with rock phosphate for its phosphorus, with potassium bearing rock for its potassium and all else which some pulverized granites contain. Since his soils were not necessarily highly productive initially, he must build that sustaining fertility in balance in the soil's reserve for a long time, as well as in balance for each season when that reserve calls for supplements as starter fertilizers to balance the fertility outflow of the season for the particular crop in question. Since virgin soils are mostly all gone, and soils are not mines any longer, this is the requirement over extensive areas if they are to grow the protein-rich crops giving health to livestock and people. It is this enormous need to rebuild sustaining fertility in eroded and in shallow, depleted surface soils by heavy and costly applications of limestone, rock phosphate, and potash rock, and then the regular seasonal use of starter fertilizer, that has not been put into the commonly considered agricultural costs. These latter were set up under the assumption that soils are perpetually productive on their own and not in terms of costs to the nation as a whole rather than to the farmer only.

This neglect of considering the costs of soil maintenance, as a food cost to everyone and not to the farmer alone in present economic thinking, is all the more serious according as the soils were initially more highly developed, which they were in the eastern and southeastern United States. Economic pressures on the farming business make it difficult now when soil mining practices must be reduced and discontinued; but management for maintenance of fertility with larger fertilizer purchases accordingly must be multiplied, but prices of farm products are not rising accordingly and cannot be set by the farmer as the creator of the products.

Our recognition of how seriously we have mined the fertility out of the soils is coming just when we have a taxation increased to the burdensome magnitude, under the erroneous assumption that the fertility of the soil is perpetual and that it costs nothing. The poor animal health and troubled reproduction—like the increased soil erosion—have been telling us that our soils have been mined. But those supposedly dumb beasts have been speaking in a foreign language to the deaf ears of their masters and in a voice that has not carried to the congested urban areas. As an economic matter, it will be increasingly disturbing for any landowner to assume the mounting costs of rebuilding the reserve mineral fertility—which we must do in the years ahead—when the consumer wants food at past prices in which he paid nothing for the exhaustion of the soil fertility which those food costs should have represented. His meats, that is all the proteins or the keystones in his nutritional support, will be more costly. Either the animal health will continue to become poorer through less fertile soil or we must contribute to the costs of healthier animals on soils restored infertility by the costly management required for that result and paid for accordingly.

These costs of rebuilding the exploited sustaining fertility must be expected, and may well take their place, for repayment from the earnings in the annual crop collections. Rebuilding the mineral reserves by applying tons of pulverized rock per acre does not interfere with cropping plans. After this part of soil fertility restoration is well started, we must rebuild also the organic matter, either as a reserve but more economically as a regular turnover in crops grown for that purpose. This will interfere with the regular cropping and with the scheme for collection of cash annually, since it will call for growing crops to be plowed back into the soil at the fertility cost in that crop and of the crop itself for that season. If we are to grow the better crops in nutritional quality guaranteeing animal health, those crops will require organic compounds in the soil contributing to the plant's creation of the items giving that higher nutritional value for the cow, and for us in feeds and foods. Such rebuilding of the organic matter will be a cost in the not-too-distant future.

Virgin soils required hundreds of years to build up their stock of organic fertility, which was so quickly and so unwittingly mined out of our soils. As yet, we have not recognized the organic reserve as the shock absorber or the constitution of the soil, giving it the capacity to stand up under our ministration of heavy dosages of salt fertilizers to it. Nor has the organic fertility been credited for much of the productive power for protein-rich, organic creations keeping our cows and us healthy and serving as the medicines for both as they become sick. This disregard of the soil organic matter will be most costly as the facts already point out in the decreasing successful use of the salt fertilizers on soils of lower natural organic matter content, and the increased use of such fertilizers only on soils higher in this natural organic fertility asset. The significance of soil organic matter will dawn, perhaps, when soils in this latter category are pulled down to a lower organic matter level, and to where the crop on the soil cannot stand up under heavier fertilizer salt treatments. If it took many years of no removal of crops from the virgin soils to build the organic matter reserve in them, can we escape the need to sacrifice a year of cropping occasionally if that growth is to maintain the soil's content of organic matter, even at a minimum for quality crop production which keeps growing our meat supply?

Agriculture that could mine the fertility out of the soil, and take over new land for little more than the cost of moving west, had little need for any economics; hence developed none of its own. But since now the declining soil fertility, growing only fattening crops, is perverting the streams of life in our domestic animals to suggest—by the increasing animal diseases and the dwarf offspring with no capacity to grow—that those streams are about dried up and headed for extinction of those animal species, that condition is now sending out a distress call for help.

We are trying to resolve the situation by economic adjustments which, to date, seem of no avail for what is a bad biotic irregularity where economics do not fit the ailment. When the farmer must buy on a seller's market and sell on a buyer's market, and when he has been paying for the loss of soil fertility in liquidations of it by that kind of economics, should he not move to the urban centers to where he would not pay those losses? The movement for that purpose and for the perpetuation of monetary capital and security in purchasable goods has been the reason for the increased urban and decreased rural population. Unfortunately, the shift cannot continue until it results in zero rural population, though still pushing hard in that direction. The farmer has been in a business that has not considered the soil, and in a distorted economics which finds agriculture with no fund for soil-saving when soils must be saved if all of us are to eat well of good meat and corresponding proteins.

UNDUE EMPHASIS ON ECONOMICS AND MECHANICS DISREGARDS THE SOILS AS THE FOUNDATION OF LIFE. WE MAY MUDDLE THROUGH

We are slow to appreciate the cow as a good symbol of the physiological requirements of all life that can be fulfilled by its being properly nourished by agriculture, only when the proper soil is properly tilled and treated for supplying—through the plants—all the elements and compounds in food for growth, protection and reproduction. Modern agriculture is threatening to put economics and the mechanical mining of the soil by more plowing so high over the cow, that even the crops are not grown for her nutrition but rather for their bulk vegetative delivery. We are putting unlimited faith in the microbial flora of her rumen for the miraculous conversion of that bulk into nutritional values. That quaking faith rests more and more on neglected soil fertility so that the cow—even when endowed by that symbiotic connection with paunch microbes—can scarcely find enough nutritional values in her feed for both her microbes and herself to survive. Machinery of all kinds, to reduce the time of our contact with things living and natural, and the economic temptations to short cut natural procedures—like feeding urea in place of vegetable proteins—seem to be conniving to have us forget the cow entirely as the symbol of the living things created by the soil and of agriculture as once a noble art.

Some twenty or more years ago, we took to plowing under prenatal pigs following the guise of supposedly proper economics. That distorted economic thinking has seemingly become such a habit, that it is moving to bury the cow too. Living things in agriculture have been forgotten in an economics built on

A little fertility uplift in cultivated soil is quickly recognized by the cow. But her defiance of the fence does not make us respect her wisdom in looking for her medicines and her health according as the soil grows it. We try to keep her on the old "permanent" pastures where the soil fertility has had no attention. (Photo by courtesy of Wexler, Agricultural Photo Library, Wallingford, Connecticut.)

industry where living matters do not enter as part and parcel in the manufacturing process. While machinery is always a helpful tool under the proper mental guidance, still the contribution by the mind dare not be too small a part in that partnership of mind and machine. Surely we must have some knowledge of the soil as nutrition if the cow and all agricultural life is not to be pushed out by the machinery that ought to serve the cow by following behind her, rather than taking such prominence as to extinguish her in our thinking of agriculture that must be a partnership—not our complete dictatorship—between animals first and the machine second in the creation of living things.

"Science has contributed so much useful knowledge to farming," says Laurence Easterbrook, a British correspondent and farmer, "that it becomes easy to forget that farming is an art also. The more we study this problem of learning the terms on which nature will accept our cooperation, the more of an art it becomes. Our great-grandfathers realized this, and our failures today are helping us to appreciate what great artists in farming those men were. Partly by trial and error, and undoubtedly by intuition through their close contact with their environment, they raised farming to its highest peaks.

"They did so largely by studying the science of balance—the interrelationship of soil, crops and animals that produced the healthy whole. Today we give that study the impressive and unattractive title of 'ecology.' The argument is that if we start with a healthy, well-balanced soil we get healthy crops. They in turn produce healthy livestock, and since our physical bodies have been grown from the fields, farming has a direct and immediate connection with human health.

"So far, then, the connection between the soil and health has been set out on lines that should satisfy both logic and common sense. No one claims that these two things can be married by some negative action—such as refraining from using chemical fertilizers—or from growing too many corn crops in succession; all depends, as our forefathers realized, on working out a positive and dynamic method of managing our land. Nor should it be expected that even if the perfect answer were found and adopted, all our physical ills would disappear. We are humans, not gods, and we all have to work out our destiny. But on the evidence, there seems to be good reason to believe that most of the human race endure a far lower standard of health than is their birthright.

"Evidence of a rather negative kind has been recorded in various experiments with animals. It may well be argued that one should not deduce too much by reasoning from the behavior of the bodies of rats, cats, and other animals under certain conditions to the behavior of human bodies, although orthodox science is ready enough to accept such evidence when carried out by orthodox scientists. But if rats do become diseased, quarrelsome, and finally unable to live when fed on a typical western diet, and if cats do develop sterility in later generations when fed on pasteurized milk, it could hardly be regarded as recommendation for that diet, and would suggest to most people that something pretty serious was wrong.

"It is easy to blame the doctors for the state of health affairs. It would be fairer to admit that they are so busily engaged in repairing the ravages which 'civilization' is wrecking on our bodies that they have not time to consider the possibilities for building our bodies from the start with less faulty materials, thus avoiding many of the breakdowns that occur.

"There is just the possibility that we may yet save ourselves, since the idea that the human stomach is a kind of petrol (gasoline) tank that will function adequately so long as sufficient food of some kind is put into it has been dying for some time. On both sides of the Atlantic the numbers are increasing of those who believe that the methods of growing our food are of the utmost importance. There are still more agricultural scientists and doctors who, a little frightened of the convictions growing in their minds, stand poised at the Rubicon between the opposing prophylactic and dynamic conceptions of health in man and the soil."

Agricultural research in soils must be challenged by some of the fundamentals that are not measured completely by criteria, including no more than yields as bushels or cash returns per acre or man hours. Qualities that deal with life, not quantities of materials alone, must be emphasized. In that research, the farmer too must share some of the thinking responsibilities. The experimenter in research may well think with, but dare not think for, the farmer. As more folks, farmers and all of us think about the fundamental processes of creation by which the soil supports all that we call agriculture, we will not be contented with the mechanics of it, or complain of the high costs and the low prices when only the speculative aspects of agriculture are considered. We will invest ourselves more in understanding the production of food for health, or our major national wealth, according to the fertility of the soil. The cow, then, as the symbol of all that is living in agriculture along with all humans dependent on it, will be convincing us that not only animal health but human health too is dependent on the fertility of the soil.

CHAPTER 14

Epilogue:
Good Horses Require Good Soils

THE QUALITY OF THE BONES determines the quality of the horse. The bones depend on breeding and the quality of the feed. The quality of the feed, in turn, depends on the efficiency with which the plant factory uses air, rainfall, and sunshine to bring bone-making minerals along with others from the soil into the vegetative bulk—all to be combined into essential and complex compounds that render effective service in the animal physiology. Thus we feed our horses and other animals accordingly as we provide fertile soils. It can be said, in truth, that good horses require good soil.

It would be presumptive to define "good horses." It might not be excessively presumptive to define the term "good soil," now that we are beginning to conserve it after recognizing that it isn't as good as it was not so many years ago. We are beginning to appreciate our changing soils as they reflect their declining fertility by an increase among animals of nutritional deficiencies commonly cataloged as diseases. Being well-fed and being healthy go together for horses as well as for humans. In case of ailing animals, we may well give emphasis to feed quality for disease prevention rather than go to the veterinarian for cure. More and more men interested particularly in the heavy consumers of forages, namely horses, cattle and sheep, are in agreement with Mr. Murray, the octogenarian Hereford breeder of Missouri, when he said, *"They aren't doing on this land what they did here 50 years ago."* These declines are reflected more boldly in the delicately adjusted body functions—particularly reproduction—that have been going lower and lower while we have been streamlining our animals for delivery of more speed, more milk, more young and a greater output of other animal services on rations consisting more and more largely of forages both green and dry. Even if we are unable to define a really good soil, we do know that our soils must be made better in fertility if forage is to take the larger share in animal rations.

DECEPTION IN FORAGES WHERE TONNAGE OBSCURES COMPOSITION

Forages may be more deceptive feeds than grains, though both may be of low efficiency. Chemical analyses of grains do not vary widely, yet the same grains are not necessarily of equally effective feeding value. Forages are subject to wider variations. The survival of the plant species in nature has been dependent on a seed composition sufficient to guarantee the plants. With constant seed composition, more fertility in the soil means more seed. Thus seed harvests have been approximate measures of the plant nutrients delivered by the soil.

When a plant can make much forage yet deliver no seed, the wide fluctuation in chemical composition of the vegetation should become evident. Grass crops that are measured in terms of tons of forage in place of seed yield per acre may be growing on soils too poor to make seed, yet we accept their forage without suspecting defective composition and poor feeding quality. Such soils have mainly a site value and serve largely as plant anchorage. It is this soil property that makes forages deceptive as feeds.

POOR SOIL, POOR PLANTS

We are beginning to realize that even though the soil contributes but 5% to the plant—mainly as its ash—the supply in the soil for that small contribution has become so low in the course of our crop juggling, in attempting to maintain tonnage yields, that we are adopting plants which manufacture their bulk less and less from what they get from the soil and more and more from what they get from air, water and sunshine. They are, consequently, more valuable as fuel and less valuable for body growth. They are more carbohydrate and less calcium, phosphate or protein. Being made mainly of wood, they cannot make good bone. They may not be synthesizing some of the more complex growth substances, like vitamins, that we know plants produce and which, in but small quantities, are so essential to life.

As we run away from declining soil fertility manifested by decreasing grain yields, and court grass cover as the cure for erosion, let us not be deceived by forages whose feeding value we still measure in tons rather than by the fertility of the soil producing them. When even the germination of the seeds we plant is apparently reduced on our lime depleted soil, as some experiments at Missouri indicate, shall we continue to be startled with the increasing manifestations of animal deficiencies? With the soil foundation of good feeding values in forage going downward, while we are streamlining our animals for more economical gains on roughages and less time from mother to market, these two contributors to effective present agricultural production are diverging so far as to invite disaster to the animal industry. Evaluation of forages in terms of their feeding value, based on soil fertility rather than in terms of tons, offers the only escape from the forage deception.

SOIL MINERALS DETERMINE FORAGE VALUES

If, in the final analysis, the soil's contribution to the plant is mainly minerals, then the more fertile soils must be fertile because they contribute more minerals. Nature has given wild animals some instincts which are demonstrated in their search and selection of feeds according to mineral supply. Calcium and phosphate as bone builders occupy first place on the animal's list of mineral dietary requirements. When additions of these two as soil treatments increase plant growth so commonly, might it not lead us to the suspicion that animals feeding on forages grown on the untreated lime-deficient and phosphate-deficient soil could be suffering bone troubles? Can the legs of the horse be clean and the bones able to take the strain when built from plants which themselves are suffering lime and phosphate shortages?

ANIMAL CHOICES AN INDEX TO THE EFFECT ON PLANTS BY SOIL MINERALS

Wild animals apparently know their medicines. Dr. A. J. Carlson reports the pregnant timber squirrel eating bones. We may well ask ourselves what becomes of the deer antlers after they have been shed in the wood? Why are rodents caught eating them? Animals practice their preventive medicine. Observers tell us that the pregnant female mule deer selects her food as carefully as the pregnant mother. Buck deer about to grow horns travel miles in search of waters rich in lime. Upland birds during mating season resort to shellfish to help build eggshells. Wild animals can roam in search of their medicine, but our domestic animals, confined by fences to fields deficient in lime and phosphate, must suffer the consequences of malnutrition. They already are demonstrating by choice of forages and grains that they can detect limed corn in the field surrounded by the unlimed; grass grown on soil which has had phosphate added, from grass grown on untreated soil; or can pick out soil-treated lespedeza and prairie hay in stacks amongst such not treated. Missouri farmers have learned that their animals are serving to test the soils to a finer degree than can be done by the soil chemist.

DIFFERENT SOILS BRING DIFFERENT TROUBLES
AND GROW ANIMALS OF DIFFERENT QUALITY

Diagnosing the numerous deficiency ailments that now are being listed for the different animals, such as pregnancy disease in sheep, acetonemia suggesting milk fever in cattle—even though they may not be near parturition and milk production, aphosphorosis with its red urine in cattle, osteomalacia or bone softness, osteophagia or bone eating, moon blindness or various ophthalmia in horses and numerous other irregularies, is not an easy task. Diagnosing those troubles in terms of the different soils in different places is much easier even if it may seem like so much "hokus pokus." Some fundamental reasoning may be in place here as a suggestion that such soil diagnosis may be a help in animal diagnosis.

Wayne Dinsmore has reported that "periodic ophthalmia, commonly called 'moon blindness,' frequently is seen in states east of the Missouri River and occasionally west thereof." In that statement he revealed the possibility of the dominance of deficiency diseases within regions of heavy rainfall, or on the humid soils, and their decrease with less rain, or in regions of arid and less leached soils. Here is a suggestion worthy of further thought and consideration as a principle of wide application.

Moving westward from Kansas City means traversing relatively narrow belts of decreasing rainfall and corresponding belts of soils that are less highly developed, less leached of lime and minerals, and bearing a greater dominance of leguminous plants in the native prairie vegetation. The protein content of wheat increases similarly from eastern to western Kansas according to studies by the federal pre-harvest surveys. Can it be a miracle, or a phenomenon without natural causes, that great herds of bison were supported on the prairie plains as the pioneers found them west of the Missouri River, while the forests of the east that presented such hardships to the pilgrims did well to supply them a few turkeys? Recall if you can, if you are old enough to have seen them, any single western brocho that had spavins, splints, or side bones among the trainloads of those hard kicking, hard bucking, and fleetfooted beasts. Then remind yourself, too, of the stamina those animals had coming from the short grass country of low rainfall, and from soils in which the limestone and other minerals were still near the surface, within reach of even the roots of these small plants.

Plant studies testify that calcium and phosphorus function in combination in plant growth, and their increase from the soil or in the crop goes with increased protein. Soils loaded with limed phosphorus produce legumes. Legumes are choice forage feeds because they are rich in protein and in minerals. Can we not see in these plant nutrient essentials taken so extensively from the soil by legumes, possibly the reason why legume bacteria can take nitrogen from the air and why legumes can produce what it takes to make animals grow? When calcium and phosphate are deficient, legumes cannot run their nitrogen-fixing factory. Perhaps they are also failing to run their vitamin factory. It may not be stretching the imagination too far to consider numerous shortages of vitamins in animals as traceable back to the soil through these channels. Perhaps these are some of the reasons why mineral-rich soils, which produce legumes to build up the soil organic matter as black humus, are also the soils that grow good livestock. It may be possible that we can look more closely to the degree of soil development, or its loss mainly in lime or calcium, as a key to our problem of wisely managing the soil for good horse or animal production.

PARTIALLY DEVELOPED SOILS ARE DEFICIENT IN PHOSPHORUS

If moon blindness and other ophthalmias are rare to the west of Missouri and are most common to the east, and if certain of these ailments represent deficiencies of vitamin A, and if they can be prevented by feeding green alfalfa, it might seem pertinent to inquire why alfalfa, such a heavy carrier of minerals, is so commonly the prescription where vitamins are needed, just as milk is such a good dietary prescription in case of most any sickness. Can the high mineral content in both have any bearing, if not directly, then indirectly? Alfalfa is also the crop making the highest demands on the soil for minerals as indicated by its many failures in the humid areas of mineral-deficient soils. Yet in the less leached soils of Kansas or Dakota, this crop starts for almost no more cost than the price of scattering the seed. As we go down the list of other legumes that are less exacting in their soil requirements—red clover, sweet clover, soybeans, lespedeza—and then through the nonlegumes demanding less from the soil, we are also going to lower choices of them, lower values as feed. They represent tonnage possibilities on less fertility. They represent more of sunshine and less of soil in their production, consequently give lower nutritional values in terms of bone or milk which demand liberally of calcium, phosphorus and protein.

When we learn to relate kinds of plants and their chemical composition, or their nutritive value, to soil development and to soil fertility supply, we may anticipate some of the deficiency troubles and avoid them. The more arid soils of Utah, for example, where alfalfa is common forage, have not developed much beyond the broken rock stage in which lime is plentiful.

These soils have not yet weathered enough to reduce the mineral bulk by removing the more soluble substance, nor to concentrate the less soluble concentrated phosphatic carriers, hence forages grown thereon will be rich in calcium, but not so rich as they should be in phosphorus. Applications of phosphatic soil treatments can then make the rainfall 32% more effective, as the late Professor Hutton of Dakota reports from 20 years of soil fertility studies in that dry region. Animals fed alfalfa hay grown on low phosphorus soils, of which the low phosphorus supply is further diluted by carbonaceous beet pulp, will not suffer calcium shortage, but instead, will suffer from aphosphoris and will deplete their body phosphorus to the point of danger, or even death, because of phosphorus deficiency in the bloodstream. If cornbelt grain supplements, which are low in calcium but high in phosphorus as well as carbohydrate content, were used with the alfalfa in place of the beet pulp, this combination should give no disastrous results as such cases illustrate; the ration depends on the fertility of the soil producing it, and may in some cases be balanced by that from other soils.

WELL DEVELOPED SOILS OF THE EAST
ARE OFTEN DEFICIENT IN CALCIUM

In the humid regions, the situation is reversed. Heavy rains have leached the soil almost free of calcium and of other bases in the plant root zone. Only the less soluble remain. Among the latter is phosphorus that has been concentrated in the soil and in the organic matter cycle of growing plants and of decaying humus residue, and even in rock phosphate beds of Kentucky and Tennessee. In these humid areas, grains are the common crops, and the legumes are the uncommon—except when soil treatments permit legume growth. Our animals given the grains will obtain phosphorus; but grains supplemented by our nonleguminous forages from lime-deficient soils mean a ration deficient in calcium and in complex plant compounds dependent on calcium for their manufacture by plants. Our soils are being depleted, however, of their organic matter so rapidly that both calcium and phosphorus are fast becoming deficiencies in forages. Animals are being pushed nearer to more of these and farther away from mainly grain rations, while we are going to pastoral farming as an antidote for soil erosion troubles.

THE SOUTH IS IN FOR MORE DEFICIENCIES

Deficiency troubles become still more acute as we add higher temperature to the high rainfall of the humid region and go to the southeastern states. Here the development of the soil has moved the original rocks so much closer to complete solution that the clay remaining contains little mineral of nutritional value, and can hold little nutrient that may come to it from solution by nature or by fertilizer. Wood vegetation is natural there, and even the legumes are of a variety refused by animals. Deficiency diseases will be rampant, and the South will become diversified only when it looks to its soil as has already been done, for example, by some of its farsighted farmers who appreciate the soil. Mr. S.L. Coleman of South Carolina, reporting in *Capper's Farmer,* says, "Thirty-five years of farming—together with 20 years of practicing veterinary surgery—have taught me that a large portion of diseases of stock can be traced to the soil. For instance, about 15 years ago, I began to have trouble with my cows aborting. Almost invariably they dropped their calves at about three months. A number of neighboring farmers had the same trouble. We were convinced that it was due to Bang's disease...

About that time I changed my farming program and began to rotate crops, planting more legumes...I began liming the land before planting peas, lespedeza and other clovers. The noticeable improvement convinced me that I had discovered the trouble...I haven't had a sick cow on my farm for ten years. I plant an abundance of cover crops and legumes—all of which I grow on land rich in calcium and phosphate. The calves are lusty fellows from birth and there has not been an abortion in the herd in years... What is true about aborting is true of milk fever also, in my experience." He continues, "I have not had a case of milk fever on my farm since I changed my farming and feed program."

In the south, with its extremely depleted soil, the animal business is already in disaster, at least when a cold snap, that barely kills beans, will kill 1,300 head of cattle as Florida experienced this past winter. In soils not so highly leached, it may not be disastrous yet, but the symptoms are giving us fair warning that if we are to feed animals successfully, we must treat the soil.

Our horses, you may say, have not yet gone so near to danger as to cause alarm. But with their longer period required to reach maturity, and with our handling less of them than of most other animals, the seriousness is minimized through lack of sufficiently impressive or extensive observation. Then, too, out of pure habit, we have given our horses so much veterinary attention by dosing with medicine, and sale barns are such a common outlet on the first suspicion of defect, that we take the route of escape rather than of reasoning, thereby failing to discover that our feeding service is defective because of declining soil fertility.

Patent feeds and mineral supplements to the ration, rather than inspection of the soil that grew it, also have become a habit. This habit has led some persons to believe that calcium and phosphorus fed directly as limestone and bone meal, in most any feed combination, are as effective as when the ration contains forages and grain from soils well-stocked in lime and phosphate to contribute these and their plant effects indirectly via the vegetative route. All are aware that oat straw is low in lime and phosphate. It is also low in protein or nitrogen.

Would you reason that oat straw supplemented with bone meal and saltpeter should be a good ration? The straw part of oats cut while green is relished by animals, but it contains at that stage of growth much that is not there when the plant is mature. Both the kinds and the amounts of chemical elements in the composition of the straw have changed while maturing. Because common salt went into the animal rations about 125 years ago to supply to animals the two elements—sodium and chlorine—needed by them, but not by plants, we are able to conclude that when deficiencies of calcium and phosphorus—the foremost soil material requisites for plants—are lowering the quality of vegetative growth to the point of substituting the wood-producing plants for the protein producers, we can raise the animals by giving them the equivalent of excelsior if only we use plenty of limestone and bone meal supplements. If we will let the animals manifest their own choice between a ration from fertile soil and the deficient ration with its mineral supplements, their judgment will render decision in favor to the crops from the fertile soil. They will select different plants, or the same plant from different spots in the same field, and demonstrate that their choice goes back to the differences in soil fertility. Poor feed plus mineral supplements must not be accepted as the equal of feeds grown on fertile soils.

The fallacy in the mineral supplement practice is the assumption that supplementing the ration with lime, or with phosphate, brings the same improvement as putting these through the plant. Such reasoning gives calcium and phosphorus put into the soil no other function than to be dragged into the plant, to occupy space there, and to be thus transported into the animal's digestive tract as they would be if shoveled from the rock pile into the ration.

**EXPERIMENTS DEMONSTRATE IMPROVED NUTRITION
THROUGH SOIL TREATMENT**

Experimental studies with sheep given lespedeza hays from limed and unlimed soils, both of which soils were given phosphates, have demonstrated the effectiveness of this soil treatment as a means of making better animals and better wool growth. Nutritional studies with rabbits show that the unlimed hay may be well digested by the animal. Analyses of the feces reveal a relatively high removal of the calcium, of the phosphorus and of the nitrogen from such hay from untreated soil in going through the digestive tract. However, these three nutrients from lespedeza grown on unlimed soils are not retained by the body. They are lost to a high degree in the urine. This loss, through the urine of these elements contributed by the soil, did not occur when the limed lespedeza was fed. Balancing the diet for the plant by means of liming as a soil treatment was the equivalent of balancing against this digestion a much more complete absorption and synthesis of these nutrients into body tissue for animal growth and profit. Liming the soil made animal feeding truly animal production rather than a case of feeding animals for their digestion of the feed and its conversion into manure. Perhaps herein is the reason why many cattle feeders look at the balance sheet at the close of their season's feeding operations and say, "Well, though I made no cash gain, I still have the manure as the profit." Few are they who can survive the economic strain of simple manure making under the guise of being beef producers.

Horses are so close to humans that these nutritional problems are more far-reaching in significance than mere economic consideration of only horses would make them. Fuller understanding of vitamin functions in humans ought to make our reasoning return better health to ourselves as well as to the animals. If vitamins are synthesized mainly by plants, it seems no stretch of fancy to imagine that vitamins might have been coming into significance through their storage, or absence, because declining soil fertility handicaps their production by plants. Some dietary work in Europe with tomatoes, carrots and potatoes with children as the test species demonstrated very favorable influences through the use of fertilizer for improved vitamin effects. Can we stand idly by to see composition of plants in calcium and phosphorus going lower and lower when not only horses but humans are potential victims of our neglect of the soil? Antidotes in the form of medicine for humans, or as mineral supplements for animals, are poor substitutes for highly nourishing vegetables, or milk, meats and forages all coming from fertile soils. We are about to realize that good health lies very near to good fertile soil. When this realization matures to its fullness, each and every one will assume his share of the responsibility in the maintenance of the fertility of our soil, and thereby the better health of both man and beast.

INDEX

A

Acetonemia, 181

Acid base balance, 71, 79

Acid soils, 14, 16, 25; important in plant, animal nutrition, 43-45; reduced by liming, 39, 43, 44, 53; reduced by magnesium carbonate, 61; *see also* hydrogen

Adsorption-exchange capacity in soils, 16, 22

Agricultural economics, 165-178

Agricultural products, *see* crop quality, crop yields.

Agriculture, traditional, 165, 166

Alfalfa, and amino acids, 151; chemical composition of, 33-36, 59; and trace elements, 127, 129, 183

Alkaline soils, 10, 16; and trace element deficiencies, 129

Amino acids, 3, 64, 91, 110, 111; essential, in clover, 123; essential, in timothy hay, 120; and nitrogen, 153; and proteins, 136-140, 142, 143, 146, 152

Anabeana cylindrica, 35

Andropogon virginicus, 91

Anemia, 164

Animal diseases, and nutrition, 2-4, 69, 70, 181-186; protection from, 6, 39, 51, 52, 62-64, 141, 149-164; *see also* animal health

Animal health, calcium essential for, 39-52; and feeds, 32-38, 81, 153, 157; and magnesium, 60, 62-64, 66, 67; and phosphorus, 69, 70, 78-80; and essential proteins, 135, 149-153; and soil fertility, 1-8, 83, 165-168, 175-178; and trace elements, 29, 120-133; *see also* animal nutrition

Animal nutrition, 2-4, 19-27, 37, 38, 69, 70, 181-186; and ash elements, 29, 130-132; and need for proteins, 141, 142, 149-164; *see also* animal health, feeds, forages, hays

Animal selection of feeds, forages, 35, 50, 81-95, 150, 154, 159, 186

Antibiotics, and disease protection, 108-112; and trace elements, 117

Aphosphorosis, 181

Apple, 129

Ash elements, 4, 19, 21; *see also* minerals, trace elements

Aspergillus niger, 115, 116

B

Australia, copper in soil and sheep wool, *xiii,* 117, 118

Bang's disease, *see* brucellosis

Barley, 129

Beans, effect of manures on, 143-146

Beef production, 6, 7; and soil fertility, 150-153; *see also* cattle

Beets, 41, 129

Beri beri, 164

Biosynthesis, *xiii,* 20, 29

Biotic pyramid, *xiii,* 4

Bison, American, distribution of, 5-7, 21, 28, 167, 182

Bloodstream, 141; and ash elements, 21, 22; and calcium, 34, 51, 52, 78; and phosphorus, 78-79

Bluestem grass, 7

Bones, 42, 55, 81; and horses, 179; and phosphorus, 71, 78, 80

Boron, 35, 44, 62, 113, 124, 127, 129

Bromine, 114

Broom sedge, 91

Brucellosis, 96, 132, 184

Buchloe dactyloides, 36

C

Cabbage, 129, 139

Calcium, essential for health, 39-52, 68; in forages, 23, 24, 26, 34; in soils, 10, 11, 15, 22, 25, 27, 101, 132, 182, 184; needed to mobilize phosphorus, 75, 76, 80

Calcium carbonate in soils, 11, 16, 44, 48, 53, 74, 75; *see also* limestone

Calcium-magnesium ratio, 53, 54, 58

Calcium-potassium ratio, 24-27, 34

Calcium-phosphorus ratio, 24, 26, 34, 77, 78

Caliche, *see* calcium carbonate

Calories, 32, 33

Carbohydrate-protein ratio, 32, 64

Carbohydates, *xii,* 20, 27; and animal nutrition, 3, 19, 22-25, 32

Carbonaceous plants, 30

Catalysts, magnesium as, 54-56; trace elements as, 115

Green manure, 88, 89
Guano fertilizers, 73
Gypsum as calcium source, 48

H

Hays, chemical composition of, 33-35, 59; nutritional value of, 47, 75, 92-94; *see also* alfalfa, clover, timothy
Histoplasmosis, 153, 156
Hogs, 62; digestive system of, 21; distribution of, 7-9; select best feeds, 87-89; *see also* pork production
Horses, effects of magnesium deficiency, 62, 63; health of and soil fertility, 179-187
Human body, bloodstream and disease resistance, 158; chemical composition of, 54
Human health, economics of, 165, 177, 178
Human nutrition, 41, 66, 187; and amino acids, 137, 138; and dental health, 155; and disease prevention, 153, 155, 156; and protein deficiency, 163, 164
Hybridization of plants, 32
Hydrogen in soils, 14-16, 25, 42, 48; *see also* acid soils

I

Idaho, native forages of, 35
Illinois, pork production in, 6, 8
Indiana, pork production in, 6, 8
Indole, 141, 143, 146
Inorganic elements, *see* ash, minerals, trace elements
Insect pests, protection from, 112
Inverse Yield-Nitrogen Law of Nature, 32
Iodine, 113, 114
Iowa, pork production in, 6, 8
Iron, 34, 113, 114, 128, 164

K

Kale, 41
Kansas, rainfall and soil in, 15; hays produced in, 47; soil fertility in, and disease resistance, 156
Kwashiorkor, 130, 164

L

Land, as economic asset, *x*, 168-171; in farms, 3, 7, 9
Legume crops, *xiii*, 81; and calcium, 39, 43, 47; and magnesium, 57-62; and nitrogen, 32; and phosphorus, 74-75
Lespedeza, *xiii*, 74-76, 186

Limestone, *xiii*, 15, 34, 76; as calcium fertilizer, 48-51, 81, 85, 186; dolomitic, 53, 60, 63; used to reduce soil acidity, 39, 43, 44, 46
Little leaf disease, 129
Liver, and trace elements, 130, 131
Livestock, distribution of, 6, 7, 9
Lysine, 138, 139, 153

M

Magnesium and calcium, 53, 54, 58; essential for health, 53-67; and potassium, 61, 62; in soils, 11, 13, 22, 25, 34, 41, 101, 126; salts of, 56, 57, 62, 148
Magnesium-carbonate, 61
Magnesium-lime element, 34
Manganese, 35, 44, 62, 113, 125, 134; deficiency of, 129, 164
Manure, as fertilizer, 102, 103, 143-147; as poultry feed supplement, 90, 91
Maps, of dental health and soil fertility, 155; histoplasmosis and soil fertility, 156; of farmland distribution, 3; of grass distribution, 2; of ground conductivity, 5; of precipitation-evaporation ratio, 11; of pork production, 8; of rainfall, 10; of soil patterns, 12, 17
Mastitis, 63
Methionine, 64, 138, 139, 151, 152
Microbes, in soil, 104, 106, 108, 141, 142, 157-159; and trace elements, 114-117
Milk fever, 43, 51, 63, 181
Minerals, deficiencies of, 6, 24, 29; and forages, 181; *see also* ash element, trace element
Minnesota, soils in, 15
Missouri, pork production in, 6, 8; rainfall and soil in, 15, 35, 85
Molds, 115, 134
Molybdenum, 113, 114, 129, 133
Moon blindness, 181-183
Mushrooms, 115, 141
Mustard greens, 41

N

New Zealand spinach, 41
Nitrogen, content of, in high bulk crops, 31, 32; excessive, in fertilizers, 84, 85, 109; nitrates, in organic matter, 106, 107, 109-112; in proteins, 136-138; in soil, *xii*, 17, 22, 25, 101, 106, 116, 169; structure of, 20, 21; *see also* protein